ENVIRONMENTAL OFFSETS

Editors: Shelley Burgin and Tor Hundloe

CSIRO

PUBLISHING

A catalogue record for this book is available from the National Library of Australia.

ISBN: 9781486313181 (pbk)
ISBN: 9781486313198 (epdf)
ISBN: 9781486313204 (epub)

How to cite:
Burgin S, Hundloe T (Eds) (2021) *Environmental Offsets*. CSIRO Publishing, Melbourne.

Published by:

CSIRO Publishing
36 Gardiner Road, Clayton VIC 3168
Private Bag 10, Clayton South VIC 3169
Australia

Telephone: +61 3 9545 8400
Email: publishing.sales@csiro.au
Website: www.publish.csiro.au

Front cover: (top) Sheep standing in front of solar panels (photo by The University of Queensland); (bottom, left to right) Black-throated Finch (photo by Eric Vanderduys), Green and Golden Bell Frog (photo by Bernard Spragg).

Set in 10.5/12 Minion & Stone Sans
Edited by Adrienne de Kretser, Righting Writing
Cover design by Cath Pirret Design
Typeset by Desktop Concepts Pty Ltd, Melbourne
Printed by Ingram Lightning Source

Contents

About the editors

Emeritus Professor Shelley Burgin was one of the first wave of female, mature-aged students to enter university in the mid-1970s. Her undergraduate degree, gained at Griffith University, was the first environmental science program of its kind in Australia. Her Master's degree was undertaken at the University of Papua New Guinea in association with the IUCN/FAO's crocodile project aimed at developing village-based crocodile farms. Her PhD was in evolutionary genetics at Macquarie University. In 2001, Shelley was appointed Fellow of the Royal Zoological Society of NSW, and Emeritus Professor (University of Western Sydney) in 2011. More recently, in 2018 she was invested as a Member (General Division), Order of Australia 'for significant service to environmental science and education as an academic, author, and mentor, and to zoology and conservation'. With close to 200 publications, including co-editorship of several books, extensive undergraduate teaching in environmental disciplines, and substantial numbers of graduate students, in semi-retirement Shelley continues to undertake research and publish.

Emeritus Professor Tor Hundloe was one of the pioneers of modern environmentalism. In 1974, he was a proto-green candidate for the Australian Parliament before there was a Green political party anywhere in the world. During the 1980s he was Director of the Institute of Applied Environmental Research at Griffith University and, during the 1990s, went on to be the Environmental Commissioner of the Industry Commission and Chair of the Wet Tropics Management Authority. In 2003, he was the first Australian recognised by the award of an Order of Australia for his development and practice of economics in line with ecological reality and ethical imperatives. More recently, he was Professor for eight years at Bond University, and in 2010 was awarded the Individual Award by the United Nations Association of Australia. Currently he is a researcher within the Global Change Institute, University of Queensland.

List of contributors

Dr Johari Amar is a Tanzanian environmental planner with a strong interest in protecting the built heritage. Her research focuses on the economic, cultural and environmental challenges in preserving heritage precincts and buildings in Tanzania.

Dr Lynne Armitage is an Associate Professor of Urban Development at Bond University. She came to this position from a background in chartered surveying, international development aid and environmental planning. One of her major research interests is in determining the value of heritage properties.

Sheron Chand is an environmental planner employed as an environmental impact assessment coordinator by the firm AECOM. She is undertaking, on a part-time basis, a Master's degree in international law at Griffith University.

Kayalvizhi Sundarraj Chandrasekar is an architect and town planner. Her research is focused on developing an urban design framework for 'smart cities' in India, her home country.

Ella Dewilde is an environmental scientist currently completing a Bachelor of Economics degree.

Keeley Hartzer is an environmental scientist employed as a Senior Environmental Officer by the Queensland government. She is also engaged in postgraduate research.

Dr Craig Langston is Professor of Construction and Facilities Management at Bond University. He has a combination of industry and academic experience spanning 40 years. His research interests include measurement of sustainable development and lifecycle analysis.

Dr Alan Midgley is an ecologist consulting both to governments and the private sector.

Dr Eric Fru Zama obtained his first degree (with Honours) at the University of Buea, Cameroon. He held various environmental positions in his home country before undertaking a Master's degree in Australia and a PhD at the University of the Chinese Academy of Science, in China.

Acknowledgements

Numerous people have played important roles in the preparation of this book. Conversations with professional staff in the Queensland Department of Environment and Science and with their counterparts in the Commonwealth government helped in clarifying government policies and procedures for biodiversity offsetting in Australia. Discussions with environmental practitioners, in the main members of the Environment Institute of Australia and New Zealand, assisted in our understanding of the day-to-day practice of biodiversity offsetting. Valued comment from Michael Mahony, Newcastle University, was provided in the development of Chapter 5 on the green and gold bell frog. We appreciate their assistance.

With regard to offsetting coal-based electricity, one of us had the privilege to visit the *Valdora* solar farm on the Sunshine Coast, and we thank the local government staff for their assistance. Discussions with Queensland University's Andrew Wilson led to an appreciation of the labour force requirements in the renewable energy industry. Andrew is overseeing the construction of the university's solar farm at Warwick. Informative conversations on employment in solar farm construction were held with officials of the Electrical Trades Union.

We must also thank the reviewers who gave their valuable time in making very helpful comments. Of course, any errors or omissions are ours.

Finally, but by no mean least, it has been a real pleasure to work with Briana Melideo, Lauren Webb and Tracey Kudis at CSIRO Publishing, and we also very much appreciate the editorial work of Adrienne de Kretser.

Preamble

Both editors of this book are environmental scientists, however, they come to this field of expertise from different disciplinary backgrounds. Shelley Burgin is first and foremost an ecologist, while Tor Hundloe is an environmental-cum-ecological economist. Different backgrounds result in different perspectives, which is evident in how they, initially at least, approach the theory and practice of environmental offsetting. This is interesting in its own right – as we illustrate. However, in the final analysis when an offset is considered for implementation and one (or more) is chosen to be applied, the editors' perspectives converge.

In a formal sense, the type of environmental offset we call *biodiversity offsetting* is generally considered to have its genesis in the USA with the enactment of the *Water Act 1970* and the entry into environmental jargon of *no net loss* and *like-for-like*.

Other types of environmental offsetting, which we include in this book, have different histories. We can note the fact that Australia has been in the vanguard – if not the pioneer – in applying and, most importantly, analysing the success or otherwise of biodiversity offsets. This explains why this book has a predominance of Australian examples.

At present few countries have fully developed – that is, legislated and put into operation – offset policies, particularly in the biodiversity category of environmental offsetting. On the other hand, carbon offsetting in its various forms is widespread globally, as is offsetting the loss of jobs and income where pro-environment decisions lead to their loss. We cover all these types of offsetting.

The countries presently operating formal biodiversity offsetting are the USA, Australia, New Zealand, the UK, some European countries and some South American countries.

Introduction to environmental offsets

Tor Hundloe and Shelley Burgin

> **'Offset' – noun: 'something that counterbalances, counteracts, or compensates** for something else; compensating equivalent' (Dictionary.com)

The types of environmental offsets

Our topic is a special type of offset – an environmental offset. The word *offset* is commonly used in a formal sense in the discipline of accounting and also, but in a different formal sense, in the field of engineering. *Offsetting* is used in a lay sense when one is referring to counterbalancing or compensating for something (of a negative nature) that is to be prohibited in the interests of something better taking place.

Our subject is *environmental offsets*. If there is a common usage of the term *offset* in environmental science, it refers to *biodiversity offsets*. However, there is another relatively common use of the offset concept in environmental matters. It has come about with the developing public interest in climate change – the notion of *carbon offsetting*. Biodiversity offsetting is in the main the interest of ecologists, while carbon offsetting is generally the interest of climatologists.

In casting our net wider to apply the offset concept to the compensation of those who have to forgo their private benefits for a greater common good, such as prohibiting an existing right to, for example, cut timber in an area to be declared a national park, we introduce a third type of offset under the environmental offset banner. This form of offsetting is the interest of economists and ecologists; if cultural values are involved, anthropologists are also an interested party. Due to the fact that in the past this type of offsetting has not been formally called offsetting in the environmental literature – rather the term *compensating* has been used – it warrants introductory, background comment.

The type of offsetting involved in saving precious natural, cultural and built environments is likely to involve compensating those who use, or have the legal right to use, these environments to provide income, jobs, food and other resources. This type of offsetting includes a range of *compensating* measures such as arranging alternative employment, providing access to other sources of resources – for example, plantation forestry to replace native forests – or monetary compensation. In developing countries, this type of offsetting takes on immense importance. For example, if tribal people are reliant on traditional bush foods and on timber for firewood, and place high cultural value on natural attributes, the question becomes whether it is possible to compensate these people if, in the interests of biodiversity protection, they lose some – possibly all – of their rights to land?

To be inclusive, we include yet another form of offsetting. This is the case when we replace a polluting method of producing a good or service by a non-polluting method. In a situation where we want or need the good or service in question – that is, we are not willing to go without it – but it is causing environmental harm, we might be able to continue to obtain the benefits by adopting an alternative process or method. Moral philosopher John Broome in his 2012 book *Climate Matters: Ethics in a Warming World* has assigned this type of offsetting the description of *preventive offsetting*. As Broome notes, these offsets mean: 'Instead of taking carbon dioxide out of the atmosphere, they make sure less gets into the atmosphere in the first place'. They prevent gas that would have been emitted from getting emitted. This is a suitable definition and should come into general use.

Preventive offsetting is best understood by reference to the production and provision of electricity. In industrialised countries, we want/need electricity in the home, offices, factories and mines and on farms. Until the fear of climate change became widespread, in most industrialised countries, and particularly in Australia, the major means of generating electricity was burning coal. If we instead produce electricity by solar panels installed on residential roofs and by large-scale solar farms and wind farms, as we are increasingly doing, we are replacing one form of electricity generation by another. We are no worse off by offsetting the polluting method with the non-polluting method. Obviously, the question of relative costs comes into play – and they are likely to differ significantly between the short term and the long term. We could include this form of offsetting under the category of carbon offsetting, discussed above, but treat it as a separate category.

There is a growing literature, most of it on a case study basis, focusing on biodiversity offsets and there is considerable documentation on carbon offsetting. In the latter case, there are numerous practical guides suggesting how to calculate one's greenhouse gas emissions from various activities, followed with advice on the means to offset the emissions. What is often missing in this literature is assessment of the success or otherwise of the specific offset arrangement. Did the relocated animals survive in what was deemed a *like-for-like* substitute environment? Was a traveller's greenhouse gas surcharge on an airline ticket actually spent on planting trees which extract an equal amount of carbon dioxide caused by their flight? Through selected case studies, we deal with a range of such matters and, where the data are available to analyse the success of an offset project, we undertake assessments.

As noted above, we include in our analysis the much-neglected *reverse offsets* (they go by different names) which are, in essence, compensating businesses and individuals when their economic and/or cultural interests are adversely affected by a decision to protect a valuable natural environment; for example, by declaring the area a national park or World Heritage Area, or otherwise restricting a prior use of the ecosystem. Our case studies from Africa and Australia make it clear that biodiversity protection would not be possible without offsetting the loss of jobs, income and natural resources that give way to environmental protection.

We also illustrate carbon offsetting by switching to a non-polluting method of electricity generation. It is a fact that we would not be in the hunt to meet globally agreed greenhouse reduction targets without offsetting coal-produced electricity by non-polluting sources.

Biodiversity offsets

Biodiversity offsets come into play when a project (development) will cause the loss of biodiversity and a government authority will only allow the development to take place if the existing biodiversity value is replaced. Of course, for a government to require an offset, it must have powers that relate to the land in question. An example would be land that is habitat for a legally defined endangered animal or plant. Projects that involve clearing large native forests are a typical example of where this situation arises. Even what might be considered minor impacts can be subject to offsetting; for example, in Queensland the removal of one or two mangrove trees to build a boat-launch ramp is likely to require an offset (State of Queensland 2015).

A like-for-like replacement is the ideal offset. The concept is no overall loss of a valued natural attribute; hence, the simplest offset is, for example, the same habitat for an endangered species; no extra greenhouse gas released after a plane journey; no loss of jobs when timber-getters are excluded from a natural area; and the same level of protection for the environment overall. No overall loss is expressed as *no net loss*.

The search for an identical replacement can lead to failure. This is particularly the case with biodiversity offsetting. To locate a replacement habitat for an endangered species is likely to be, at least, very difficult. We will involve a case-by-case assessment. If animals are to be relocated, a prerequisite is evidence of success in previous translocations – otherwise we are engaged in an experiment which might fail.

If the project developer is fortunate, there may be a nearby, or neighbouring, similar-sized area with equal biodiversity values as the area the developer will significantly modify or degrade. If the potential substitute area is in private ownership and its owner has the legal right to do as they please with the land, it is likely to be of substantial monetary value to that person. In this situation, the project developer would have two options. One is to purchase the land and hand it over to a government which would assign the land a form of protected status, say, declare it a national park. The other is to enter into an agreement with the owner of the substitute land, requiring the latter to take whatever steps are necessary to render the land a like-for-like substitute, on the payment of monetary compensation. An example would be an agreement to significantly decrease a grazing herd so that particular grasses, which are food for a flock of birds to be displaced by the development, can again grow and produce food for the birds.

Bio-banking (biodiversity banking)

A fortuitous situation such as that described above will be rare. With this in mind, *bio-banking* is promoted by governments. This concept means that someone with an eye to the future acquires and holds an area of land which has the potential to become an offset to replace a similar area that is likely to be destroyed or degraded in the future. In economic jargon, the *banked* area has monetary worth as *option value*. This is the value of saving it or otherwise retaining the option to use it as a replacement area in the future when a developer needs to offset the destruction of biodiversity values on similar land.

While the bio-banking concept is theoretically appealing, in practice it has gained little traction. This is because private landholders would need a very good crystal ball to be confident that their land will be required as an offset in the future. It is a different case if governments hold (bank) land which they have identified will be required in the future for infrastructure such as roads, railway lines and school complexes. Because governments plan these types of developments well in advance, a crystal ball is not needed.

Re-creating nature and time lags

The situations described above are not the only possibilities to offset biodiversity loss. Other potential solutions are likely to be considered. None are ideal. One is to re-create an environment similar to the one to be degraded. For example, if a forest is to be felled so that a development can occur on that land, a nearby degraded area (let us assume it exists) could be planted with similar trees. This is the example used by the Queensland government to illustrate replacing a lost environment.

In using this example, the Queensland government uses a drawing of trees being replanted near the trees being knocked down for a development. In the drawing, the new trees are shown as mature. However, even the least bio-diverse, new-forest ecosystem will take a decade to approach maturity. Here, we are thinking of a stand of fast-growing *Acacia* (wattles). Attempting to replace a mature rainforest takes hundreds of years. Even leaving the trees out of consideration, what happens to the fauna which lived in the felled forest home? What of the range of other plants that grew in association with the former trees? These questions cannot be overlooked – in fact, they are the fundamental ones. This is not to denigrate the example, which is simply an illustration, but rather to draw attention to one issue in particular – time lags in offsetting losses and the potential for irreversible losses.

The significance of time lags is a case-by-case matter. The destruction of a handful of trees common to an area is completely different from the bulldozing of hundreds of hectares of rainforest. In the former situation, the fact that the handful of replacement seedlings will take years to reach maturity is unlikely to matter in the grand scheme of protecting biodiversity. It is not so with the loss of the rainforest.

Monetary offsetting of biodiversity loss: Is this realistic? Or ethical?

The final offset option – should it be called an offset? – is a requirement that a developer who has been unable to furnish a like-for-like offset puts money into a trust fund. This money is subsequently spent in a way that compensates for the destruction of a natural environment (or heritage building or site).

This is the least preferred option. It puts the onus on government to attempt to achieve what the developer was not able to achieve. It should be the last resort, assuming it is permitted in the suite of offsetting options. However, when registers of offsets are scrutinised, it seems to be the favourite option of developers. This suggests that approval for this option is too easy to obtain. It also suggests that the required financial sums are too low. The method of determining the quantum of the monetary sum to be paid is, if not a black box, insufficiently transparent. The formulae are spelled out, but the assumptions underpinning them are vague and contestable.

The test of the efficacy of paying money is its success, however measured, in maintaining biodiversity; in other words, meeting no net loss and like-for-like outcomes. There is a tendency for governments to allocate some of the monetary contributions to what are termed *indirect* projects, such as research. When this happens, for various reasons it is very difficult, if not impossible, to gauge the outcome. The most obvious one is not knowing beforehand the results of the research. It goes without saying that much research fails to come up with useful results. That is the nature of research.

The compensating project offset

In introducing compensating project offsets, it helps to recall a little history and understand a fundamental economic theory. The concept of this form of environmental offset started to take hold not long after the publication in 1987 of *Our Common Future*, otherwise known as the *Brundtland Report,* named after the Chair of the UN World Commission on Environment and Development, Gro Harlem Brundtland. The notion of sustainable development was formulated in that publication.

The key practical goal of sustainable development is to marry economics and the environment, with the ultimate aim of maintaining a healthy functioning environment, such that future generations are no worse off than the present generation, while seeking a vast improvement in the material wellbeing of the planet's poor. These goals are called *inter-generational equity* and *intra-generational equity,* respectively. The latter means that the world's poor (most of humanity) has to be allowed – in fact, encouraged and supported – to achieve sustainable economic growth. This is an ethical stance of overall importance. However, it has to be achieved in a manner consistent with the long-term ecological health of the planet.

To achieve both of these apparently conflicting goals – environmental protection and economic growth for the world's poor – economic growth has to be defined differently from the conventional measures, such as Gross Domestic Product and its variants. Growth has to be net of environmental damage, while recognising that the material condition of humankind cannot be improved without altering the environment.

Recognising this has led economists to seek practical means of permitting a certain amount of environmental damage in seeking economic growth, while compensating for that damage by requiring actions that are environmentally positive; that is, that compensate for the damage. Economists came to use the term the *compensating project* to define an offset. The compensation is required to be of such a magnitude that there would be no net loss in the *productivity* of the environment. Productivity implies the ability to produce something, for example, we talk of the productivity of agricultural soils. But the concept is extended well beyond such obvious forms of productivity, to encompass the notion that beautiful landscapes and underwater-scapes, such as the Great Barrier Reef, produce value; that is, are productive. Furthermore, there is no productivity without the life-support functions of global environment.

To give one real-world example of an offset that maintains productivity, think of a major port to be built that would permit poor farmers in a developing country to become richer (that is, improve their impoverished living standards) by exporting products to wealthy nations. Construction of the port will cause the destruction of hectares of mangroves. Those developing the port would be required to replant a mangrove forest in a nearby degraded coastal area, with the aim of continuing the productivity of a fishery which is dependent on mangroves. This is what came to be called like-for-like offsetting. Of course, as noted above, it is not as simple as this – many years have to pass before the planted mangrove forest is capable of contributing to fishery productivity. That stated, this example illustrates that the material needs of the poor will only be met with some environmental damage and it is imperative that we seek means of compensating for that.

Cost–benefit analysis

The notion of compensating those who suffer a loss from a project has a long history in economics, not necessarily tied to environmental policies. However, from 1920 when economist Arthur Pigou published *The Economics of Welfare*, we were alerted to the fact that non-priced side effects from developments were a cost to society, called *externalities* by economists. If externalities resulting from a project outweighed the benefits, as measured by market prices (or surrogates for market prices), the project could not be accepted as meeting economic criteria. This understanding became the principle underpinning cost–benefit analysis (CBA), otherwise known as *applied welfare economics*.

The underlying principle of CBA is what economics call *Pareto efficiency*. Wilfred Pareto lived from the second half of the 19th century until 1923 (1848–1923). However, it was only in 1939 that two famous economists, John Hicks and Nicholas Kaldor, came to express Pareto's criterion as a *Pareto improvement*. In plain English, this means if a project, say, the construction of a dam makes just one person better off and no one worse off, it has met CBA criteria. Of course, such a narrow rule was virtually impossible to meet and was no use in making decisions. The rule was therefore modified to: if the beneficiaries could compensate the losers, the development would be approved. What this did not require was that the beneficiaries actually compensated the losers. That compensation of some form had to be made was the next step. This is how economists came to the idea of offsetting.

In practice, it was not until the 1950s that CBA commenced to be used although, from the 1930s the US Corps of Engineers recommended a basic CBA approach to evaluate projects. The *Flood Control Act 1939* is a reasonable explicit example of this. Otto Eckstein is considered the pioneer of CBA.

A publication in 1966 proved to be a catalyst in the evolution of economic offsets into environmental offsets. Marion Clawson and Jack Knetsch wrote *Economics of Outdoor Recreation* in which they formulated a means of determining a surrogate monetary value for unpriced (free to enter) recreation sites. In Australia, where major conflicts erupted over proposals to have very valuable biodiversity and cultural sites made World Heritage properties, this concept proved extremely important in estimating amenity, ecotourism and cultural value. Australia now has more World Heritage properties proclaimed for their natural values than any other country. This achievement hinged on economic compensation – offsetting the loss of jobs and income of the industries that gave way to nature conservation.

As economists took up the challenge of undertaking CBA on a wide range of projects, only some of which focused on environmental impacts, the US government enacted a paradigm-changing piece of legislation in 1969, the *National Environmental Policy Act*. It introduced the concept of Environmental Impact Statements (EISs). A fundamental feature of an environmental impact assessment was the weighing of a project's costs and benefits, both widely defined. The EIS concept spread across the world, along with the economists' tool of CBA.

CBA-based environmental offsets come to the fore in Australia

We will use Australian examples to illustrate the role of compensation in resolving environmental disputes. Australian environmental disputes – in some cases, on-the-ground physical conflicts between pro-environment groups and pro-development interests – have in the main been about stopping an environmentally degrading resource use and preserving an area for nature conservation purposes. They include saving the Great Barrier Reef from oil drilling and limestone mining, protecting the wilderness forests of south-west Tasmania from hydro-electricity dams and timber-getting, saving Fraser Island from the mining of rutile, zircon and ilmenite and timber-getting, and saving the North Queensland rainforests (known as the Wet Tropics) from timber-logging. In all these cases, pro-environment decisions would have a negative impact on existing industries. Only with the provision of compensation – pro-environment offsets – could governments politically justify these places becoming World Heritage Areas and thereby excluding the existing industries.

This is where the ability to measure the economic value of nature-based tourism and its stricter sister, ecotourism, came into play. Would these uses generate more income and jobs than the economic sectors they replaced? The pro-environment case was helped by the rapid increase of nature-based tourism that commenced in the late 1960s/early 1970s.

The data to measure the monetary value of tourism had to be gathered, which meant that researchers had to visit these areas, counting and interviewing tourists. Next, the monetary value that tourists put on their nature-based experience had to be calculated using the method developed by Clawson and Knetsch (1966). Finally, forecasts of tourism growth had to be made.

In all these cases, it was found that nature-based tourism would compensate for the loss of income and employment that would result if the existing industries ceased. While that result would have satisfied CBA criteria, it did not do anything for the workers who were to lose their jobs as, say, timber-getters. Timber-getters were not necessarily going to become ecotourism guides or waiters in resort dining rooms! This is where specific offsets had to be designed and implemented. The main way this was addressed in the Wet Tropics case was to offer retraining and financial incentives to workers to relocate and find alternative employment; provide money to nearby towns and regions to engage in small infrastructure projects; and provide financial assistance to the owners of timber mills. An example of the latter was payment to a timber mill to re-tool so that it could mill planation softwood instead of rainforest hardwoods.

Over time, the language of economists has changed and today a compensating project is most likely called an offset.

Environmental offsets as a tool for conservation

The precursor to consideration of the concept of environmental offsets as a tool for conservation must be the realisation that species and ecosystems matter, that they are being lost due to human activities and that it is imperative to find alternatives to stay the continued erosion of biodiversity.

A historical perspective will help us understand the path to biodiversity offsetting, commencing with our attitudes to land – particularly why and for what purpose we value land. As nature has succumbed to development due to the growth of human population (and the need to feed, house and clothe that population), we have changed our attitudes. One of the significant changes has been the protection of areas as national parks. This, of course, constrains the protection of biodiversity to relatively small areas of the planet. Much more is required and this leads to environmental offsetting.

Once there was scrub

Throughout Australian European history, removing trees from the landscape was considered good management. This activity was encouraged by successive governments, commencing with the first colonial government. Only in recent decades have some – not all

– governments acted to stop, or at least limit, land clearing in rural areas of Australia. Overlooked is the ongoing clearing of peri-urban land as Australian cities continue to expand, driven by population growth that resulted in over 82% of the Australian population living in urban areas by 2010, a further 15% in peri-urban and exurban areas while only 3% live in rural areas (Sutton *et al.* 2010). However, clearing has by no means been restricted to such areas.

Brigalow Scrub, dominated by *Acacia harpophylla*, was described by Patrick O'Shanassy as 'cheerless waste' in his publication *Contributions to the Flora of Queensland* (cited in Hutton and Conners 1999). It continued to be considered as such until recently. It is estimated that before these Brigalow-dominated ecological communities began to be cleared for agriculture soon after European arrival in Australia, they extended from Townsville in north Queensland to Narrabri in northern New South Wales (McAlpine and Seabrook undated). In this area of ~37 million hectares, Brigalow communities covered 7.5 million hectares.

These communities are now considered 'among Australia's most significant biodiversity hotspots' (Seabrook *et al.* 2016). However, most of the Brigalow (almost 90%) has been destroyed by a combination of clearing (predominantly for agriculture) and, between 1880 and 1934, by the influx of the feral prickly pear (predominantly *Opuntia stricta*) which ultimately spread to cover 24 million hectares of Queensland before biological control was successful (McAlpine and Seabrook undated). Prickly pear could out-compete Brigalow communities.

In the early days, clearing of Brigalow ecosystems proved difficult. This was partly due to the hardness of the timber which made it difficult to fell. Even more resistant to destruction were the roots. Once disturbed, the roots sucker densely and grow so profusely that within 30–50 years, if abandoned after initial attempts at clearing, complex Brigalow ecosystems may re-form (Seabrook *et al.* 2016). However, with the biological control of prickly pear and the 1946 introduction of technologies that mechanised the removal of Brigalow, the efficient removal of Brigalow ecosystems was underway (McAlpine and Seabrook undated). Mechanised removal involved a large tractor or equally strong mobile machinery pulling a massive ball and chain, which knocked the Brigalow trees over and yanked them from the ground.

By 1954, mechanised forest clearing was commonplace. In the 1960s, land clearing took place on a grand scale due to the perfect storm of improved mechanisation and government incentives for agriculture development, together with research into the control of Brigalow. Consequently, by the 1970s the Brigalow scrub had 'disappeared … [and the] battle to clear Brigalow had been won' (McAlpine and Seabrook undated). Approximately 10% of the former Brigalow ecosystems are still extant, typically as small remnants (Seabrook *et al.* 2016).

Even if considered cheerless waste by some, these ecosystems are habitat to many taxa including bush turkeys and other birds, as well as reptiles, cicadas, butterflies and other insects. However, the Brigalow scrub of Australia was not the only wilderness that was disappearing.

A Sand County Almanac: publicising human damage

Aldo Leopold published *A Sand County Almanac* in 1949, following many years of observation of the US landscape (Leopold 1990). In millions of readers around the world, the *Almanac* prompted a recognition that the world's ecosystems were rapidly changing due to human intervention.

Leopold's book continues to grow in popularity. Indeed, due largely to this *Almanac* (although he published much more) Leopold is considered one of the pioneers of modern conservation science, policy and ethics. Notwithstanding that, the book does not directly address the restoration of natural areas, and does not consider anything that could be considered as environmental offsetting. It does, however, challenge readers to protect native ecosystems and hints at the potential resilience of some species. The recognition that native ecosystems and their inhabitants are worthy of protection from unnecessary destruction is the first step towards their conservation and any consideration of environmental offsetting.

Evolution of recognition of Australian biodiversity

One early enthusiast of Australia's native flora, albeit for its perceived economic value, was O'Shanassy. In the same book as he described Brigalow Scrub as cheerless waste, he also waxed lyrical about the wilderness surrounding him in what is now known as Central Queensland, albeit as encouragement for colonists to consider it 'replete with vast numbers of economic and industrial plants' (Hutton and Conners 1999).

In the early days (late 1800s) of developing environmental awareness in Australia, most of the battles to reserve areas for wildlife were waged by acclimatisation societies (e.g. Victorian Acclimatisation Society – Phillip Island; Royal Zoological Society of New South Wales – Sydney). However, as implied by the term *acclimatisation*, the efforts of such societies and of like-minded people were to introduce and acclimatise imported species or, at best, species that were not native to the area, with the view of eventually releasing them into the wild in the hope they would survive. Unfortunately, the releases often left a legacy that underpins some of Australia's major problems with non-native species that have come to dominate local ecosystems. The case of introduced animals is well known – we still struggle to manage rabbit and fox populations – to say nothing of the huge number of feral plants.

Protecting nature: the first national parks

What we came to term the *National Park Movement* originated in successful campaigns in the USA. The world's first national park, Yellowstone National Park, was declared in 1872. More than 100 years later, in 1978, it became a World Heritage site, among a small group listed by UNESCO in the first year of such designation. Other very well-known listings in that year were the Galapagos Islands (Ecuador) and L'Anse aux Meadows National Historic Site (Canada).

Yellowstone deserves a brief description. It sits in three US states, Wyoming, Montana and Idaho. The first human inhabitants of the area, Native Americans, were there ~11 000 years ago. Other inhabitants included grizzly bears, bison, elk and wolves, plus many more animals that remain in the Yellowstone area to this very day. The first non-Native Americans to roam and settle in the area were so-called *mountain men*, many of whom were trappers seeking animal pelts for the fur trade. Today, Yellowstone National Park is a highly visited part of the USA. The famous geyser named Old Faithful fascinates many visitors, as do the native animals inhabiting the massive sub-alpine forests.

Gazettal of the first national park in Australia (subsequently named Royal National Park, situated on the southern outskirts of Sydney, New South Wales), and the second in the world, occurred in 1879. This was years before John Muir, founder of the Sierra Club and known as the Father of National Parks, was successful in his campaign to have

Yosemite National Park gazetted in 1890. As an aside, we can note the 'Royal' was added to the name a year after Queen Elizabeth II visited Australia in 1954.

In this early era, the rationale for declaring national parks was two-fold: first, protection of the natural environment, which today we would recognise as protecting biodiversity; and second, providing places of natural beauty and tranquillity where people can escape from the common chore of work. The rationale has not changed – it remains relevant today.

The declaration of Royal National Park has been attributed to lobbying by acclimatisation advocates, in this case by the Royal Zoological Society of New South Wales. Indeed, when that organisation was formed in 1879, its goals were centred on the introduction and acclimatisation of non-native birds and animals (Hutton and Conners 1999). Deer released into Royal National Park in the name of acclimatisation continue to prosper, causing ongoing conflict (Burgin *et al.* 2015).

Between 1890 and 1920, the actions of the Royal Zoological Society resulted in conflict between Royal National Park trustees and conservationists, but by 1917 the protection of native animals had been elevated to be the second aim of the Society. After World War II, the Royal Zoological Society evolved from its original focus on acclimatisation of non-native species to instead focus on nature conservation, attracting high-profile conservationists and eminent scientists to its governing body (Hutton and Conners 1999).

A major innovation of the late 20th century was the creation of World Heritage listing, which recognised that some areas, natural, historical and/or cultural, were the treasures of all humans. It is a tragedy that in ethic-cum-religious wars or lesser conflicts, some world treasures have been destroyed in the name of sectarian claims by those who refuse to recognise our common humanity.

From national parks to broad-brush environmental protection

Notwithstanding the initial success of the national parks movement in Australia, and proto-ecotourism initiatives in the late 1920s and early 1930s by folk such as Bernard O'Reilly, Arthur Groom and Romeo Lahey (who built lodges in what is now the World Heritage Gondwana Rainforests), it was nearly 100 years after the gazettal of Royal National Park that strong environmental protection emerged in Australia.

The publication in 1962 of Rachel Carson's *Silent Spring* is considered by many as the genesis of modern environmentalism. The first Earth Day, which some celebrate as the birth of environmentalism, was held in the USA on 22 April 1970. It was inspired by the protest movements of the 1960s, and as the reasons for those protests were resolved or were on the cusp of resolution, environmentalism began to move to centre stage. The concept spread around the world. It matters little which day you pick – the lessons from the protest movements of the 1960s helped fashion the environmental movement.

The emergence of the environmental movement was undoubtedly in response to such catastrophes as the 1967 *Torrey Canyon* oil spill which happened when the vessel crashed off the English coast. The resulting oil spill led to huge numbers of birds suffocating on English and French coasts. Photographs of seabirds coated in oil horrified people around the world. Two years later, there was the Santa Barbara accident – an oil drilling platform exploded and dumped huge quantities of oil, much of which washed on to the coast. Again, birds died when they became coated in oil.

In the late 1960s, a significant number of environmental books were published, some prophesying doomsday. The best known was Paul Ehrlich's *The Population Bomb*, released in 1968. On 18 March of that year, US Attorney-General and Democrat nominee for the US presidency, Robert Kennedy, made a famous speech outlining the flaws in economic accounting, particularly the inclusion of pollution clean-up costs as a benefit.

Worldwide, cities were choked with smog. There was destruction of rainforests, waterway pollution and an emerging concern for the greenhouse effect (i.e. climate change). These were the catalysts for an increasingly strong environmental movement, although the specific issues differed among countries. Australia stood out in its campaign to protect a large number of world-renowned natural environments, the Great Barrier Reef taking pride of place.

By the 1990s, concern for the environment had become a significant social and political force, although there was no consensus on the future management of natural areas. For example, Robertson (1993) suggested that environmental concerns remained very divisive, and suggested that environmental protection was the antithesis of development. Typically, the debate centred on the balance between environmental protection and traditional economic pursuits such as mining, timber-getting and constructing dams. In the 1990s, it remained difficult to integrate the apparently opposing values of the economy and the environment. However, the concept of sustainable development was being promoted. This concept came to be called ecological sustainable development in Australia to make the point that sustainable development depended on maintaining a healthy planetary ecosystem.

Offsets to the rescue

One way of resolving the differences (or at least minimising the impacts) between environmental protection and economic interests, when the latter are deemed necessary for human well-being, is through environmental offsets. Offsets are also known as set-asides, compensatory habitat and mitigation banks (Gibbons and Lindenmayer 2007). Whatever they are called, they all refer to voluntary or government-required conservation activities designed to 'offset residual, unavoidable damage to biodiversity' due to activities associated with development (Burgin 2008). Offsets were not developed to be a panacea for inferior environmental management, but rather as a tool to be used in appropriate circumstances.

In Australia, biodiversity offsets were first formally recognised with the introduction of the New South Wales *Threatened Species Conservation Amendment (Biodiversity Banking Bill) 2006*. This legislation sought to reduce the loss of ecosystems in coastal ribbon development and in development of satellite towns beyond Sydney, as well as to enhance protection of the last remaining native vegetation on the Cumberland Plain in Western Sydney (Burgin 2008). Subsequently, there have been changes to this legislation to strengthen it.

The use of environmental offsets has become accepted across the spectrum of those concerned with both conservation and development (ICMM 2005; Burgin 2008). However, there are dissenters, from both the pro-development and environment protection sides. As a tool for conservation, offsets remain controversial (Bull *et al.* 2013). This will become obvious in some of the case studies we report.

About this book

The following chapters show that environmental offsetting is applied to very different problems and in quite different circumstances. This section presents a road map for the rest of the book.

Offsetting is commonly used when, by law (generally legislation pertaining to environmental impact assessments but there are other circumstances which differ country to country), there is a requirement of developers to deal with residual environmental degradation after all attempts at mitigation have been exhausted. This is usually the case for biodiversity offsetting, however, as we explained, this is only one type of environmental offsetting.

In Chapter 2 we open by drawing attention to some of the most important issues in environmental offsetting. Some issues are contentious, some are simply difficult. Given this book's significant focus on biodiversity offsetting, we conclude the chapter with a detailed description of this type of offsetting.

In Chapter 3 we discuss in some detail the nuts and bolts of biodiversity offsetting.

Chapter 4 makes the case for the importance of declaring protected areas, such as national parks and World Heritage areas, if we are serious about maintaining biodiversity. However, we cannot expect government-sanctioned protected areas to carry all the burden – in fact, they will always remain a small percentage of the landscape. This means that off-reserve protection will continue to be essential. This is the opening for offsets to play a very important role.

Chapter 5 is the first in a series of case studies. The first tells a well-known story, although not in detail. It describes the story of Australia's best known, and very early, attempt to offset the loss of green and golden bell frog habitat as land was prepared for the Olympic Games held in Sydney in 2000.

Chapter 6 is another case study of the outskirts of Sydney, focusing on a river and recreational fishing. A range of modifications to the river has been blamed, with some justification, for reduced fish catches. How should we compensate fishers?

In Chapter 7, our focus is how – and whether – we can offset the loss of nature due to ever spreading cities. More than half the world's population live in cities. Cities will continue to expand as more people move into them and, most significantly, as world population grows from the current 8 billion to 10 billion in under 30 years. Can we develop green cities as a small step to partially offset the biological diversity that is lost as land is cleared for 'little boxes' and 'tar and cement'? Those terms were used as early as the 1960s, in the popular songs 'Little Boxes' by Malvina Reynolds and 'Tar and Cement' by Verdelle Smith.

The next three chapters go beyond biodiversity offsetting into the wider area of compensating projects and preventive offsets. Chapter 8 deals with how to offset the use of electricity. Trees can be planted to sequester carbon dioxide. However, to be effective at a large scale, immense forests would have to be grown. Better, we suggest, to change the source of electricity – in other words, seek preventive offsets.

In Chapter 9 we introduce the trials and tribulations of environmental management in developing countries, where there is an imperative to enhance living standards while protecting both natural resources and valuable built heritage. In this case, we are dealing with compensating those who give way for the common good. This warrants elaboration.

One of the two most important sustainable development matters at both a country level and a global scale is to determine the means by which developing countries can achieve a much-improved lifestyle for their people while protecting their environment as a life-support system, as well as being agriculturally and industrially productive. The equally if not more important global matter is for rich industrial countries to manage both their production and consumption on a globally sustainable basis. That is too broad an issue for this book; our focus is on offsetting in developing countries.

To illustrate some of the issues of environmental protection in poor, developing countries we have selected two very difficult case studies, one dealing with nature conservation

(Chapter 10) and the other with built heritage conservation (Chapter 11). In both, compensating those who are required to give way is imperative if conservation is to be successfully pursued. In this context, we refer to 'poor countries' rather than using the term 'developing countries'. Although some may consider the use of this term politically incorrect, 'poor' is commonly used (e.g. by *The Economist*) and, in some circumstances, 'poor' is more correct than 'developing'. This is because there has been an acceptance that some of the poorest countries are not developing and hence the language is misleading while 'poor' is a more accurate descriptor.

Chapter 12 covers all major types of offsetting, focusing on the proposed Adani coal mine in Queensland. The degradation of land by a large open-cut mine will have potentially serious impacts for an endangered species, in addition to other difficult-to-solve biodiversity impacts. Biodiversity offsetting is one form of offsetting to be explored in this context. There are related complexities – burning of the mined coal will add to the carbon dioxide in the atmosphere. How is this to be offset, if at all? The third offsetting matter is that the coal is a non-renewable resource. Should we not compensate future generations for its depletion?

Finally, Chapter 13 deals with offsetting job losses in the transition of energy production from coal-mining and -burning to renewable, carbon dioxide-free sources. This has become one of the crucial environmental-cum-economic issues of the 21st century. Australia, as a major coal-mining and coal-exporting country, is faced with pioneering policies to deal with this matter.

Mary Robinson, former President of Ireland, in her 2018 book *Climate Justice: Hope, Resilience and the Fight for a Sustainable Future* reminds us that:

> … *as we make the transition to cleaner energy, we must remember the millions of fossil fuel workers around the world who have spent their lives extracting the fuel that has fed our economies. They are victims of climate change and deserve to be treated with dignity.*

Environmental Offsets will stand as both a historical source in our understanding of environmental offsetting and as a guide to the way forward. It will illustrate what works, what does not and what can be improved.

Crucial and contentious issues in addressing offsets

Tor Hundloe

In this chapter, we focus on the crucial, and some of the more contentious, issues in environmental offsetting. Offsetting in its various forms can be contentious due to political disagreements, as is the case of the transition to solar and wind energy from coal. Some offset proposals are contentious in that a paradigm shift is required if they are to be implemented; for example, compensating future generations for the depletion of non-renewable resources, such as minerals in the ground. When it comes to the increasing use of biodiversity offsets there are complications, which in practice are possibly beyond resolution. For this reason, we pay considerable attention to the theory of biodiversity offsetting.

Carbon offsetting

Carbon offsetting can take various forms. We do not discuss all of these in detail, as carbon offsetting is a book-length subject in its own right. We briefly mention issues that we do not deal with further in this book, then move to our focus on the difficulties of matters which are covered in this book.

A subject not given a chapter in the book is the role of individuals in carbon offsetting. Some individuals engage in voluntary carbon offsetting when making car trips, for example, by giving a donation to an organisation which will plant a tree (similar to offsetting on flights). Another example is the installation of energy-efficient household whitegoods and lighting. Increases in efficiency in electricity use will be reflected in lower household power bills, an individual benefit; the decrease in carbon dioxide emissions is a public benefit.

There is information available by which individuals can calculate the amount of carbon dioxide emitted in a journey, in the choice of a whitegood, or in various lifestyle choices. Such calculations allow individuals to ascertain the amount of carbon dioxide offset needed to live a carbon-neutral life.

A popular means of offsetting carbon emissions in the household sector is to install solar panels; that is, to replace coal-fired electricity delivered via a grid with electricity formed chemically in roof-top photovoltaic panels. In fact, Australia leads the world on a per capita basis in the installation of roof-top solar panels – one in five homes has installed them. Queensland, the Sunshine State, is the front-runner.

It is not only householders who engage in offsetting the carbon emissions from burning fossil fuels to produce electricity. A large local government authority in south-east Queensland, the Sunshine Coast Council, with a residential population larger than most Australian regional cities, has constructed a solar farm capable of producing enough electricity to meet all the council's needs (administration offices, libraries, public swimming pools and the like). The solar farm, known as Valdora, commenced producing electricity in mid-2017. Another example of large-scale voluntary carbon offsetting is that undertaken by the University of Queensland, which aims to be the first university in the world to be carbon-neutral. It has a small carbon farm on its Gatton campus and is building a large one on farmland on the outskirts of the rural town of Warwick.

Both the Sunshine Coast and University of Queensland initiatives involved controversy over siting of the solar farms. In the former case, glare was the issue, while in the latter it was the covering of farm land. In the former case, the issue was resolved by planting trees on the border of the farm, and in the second it was made clear that the land itself would not be degraded, and sheep could graze around the rows of panels. Chapter 8 discusses the Valdora solar farm in more detail.

The rapid increase in renewable energy, solar farms, wind farms and pumped-storage hydro-electricity is the centre of a long-running and continuing debate on energy policy in Australia, and all forms of environmental offsetting will be in play.

The energy argument

In Australia, there are constant arguments among the political parties over the need to maintain coal-fired power stations due to their reliable production of electricity, contrasting this with the variability of supply from solar and wind farms. There have been failures of supply by renewable sources, due to extreme weather events. Some of the controversy is based on mistaken premises, though. For example, in South Australia many comments were made about the unreliability of supply from wind farms as a factor in the blackouts of 2016–17. However, those failures were due to lightning strikes that resulted in trees falling onto power lines. The blackouts would have occurred no matter whether the power came from renewable or coal-fired electricity sources.

Despite the continuing political arguments over the most reliable form of energy, South Australia is on track to be generating 75% of its power from renewables by 2025. Elsewhere in Australia, the community is increasingly embracing renewable forms of energy. The spectacular fall in the costs of solar, wind and battery power is driving the switch as much as the concern about climate change. Non-polluting electricity supply, driven directly and indirectly by the sun, will increasingly offset the impacts of atmospheric pollution which would have occurred had the electricity been provided by the burning of fossil fuels.

Compensating for the loss of jobs and profits

Another important form of environmental offsetting, although not widely recognised as offsetting, is decisions by governments to provide environmental benefits, usually by expanding the conservation estate. This concept was introduced in Chapter 1, and here we will elaborate. Expanding the conservation estate is done by removing activities that are damaging it. A common activity in forested areas, particularly rainforest areas, is timber-getting. To remove this industry from the forests, governments need to compensate the businesses and workers who are put out of business and work.

This practical pro-conservation action is generally supported by the community as a whole. However, such action can be bitterly fought by those who do not want to relinquish their rights to use the environment in the manner they previously had. It is accepted as fair that people should not lose their income for the greater social benefit.

Most of Australia's World Heritage areas have been acquired by the process of compensating those who have had to give way. Rainforest areas in north Queensland, south-east Queensland into northern New South Wales and south-west Tasmania were being logged before they became protected areas. Fraser Island was being logged as well as being mined for rutile, zircon and ilmenite. In all these cases the loss of jobs was offset by giving assistance to workers so that they could transition to other employment. Businesses were compensated in various ways.

It is unlikely that readers do not know of these Australian examples as all of these now-protected environments have become international tourism drawcards. However, we are unlikely to have knowledge of this form of offsetting in very poor countries. If these countries wish to protect natural areas such as rainforests, they are likely to take land from traditional food production or otherwise impact the lifestyle of tribal people. Helping tribal peoples to transition requires a form of compensation that focuses on the loss of lifestyle. This is no easy task. While humans can be relocated, often easier than animals, most people do not take to this lightly. How do we replicate a tribal lifestyle if the same natural environment from which they have been moved does not exist elsewhere? Yet, the common good of a poor, developing nation is likely to benefit from protecting and promoting its natural heritage. Nature-based and ecotourism can be very profitable, and provide employment for guides. It is possible that some displaced people can find rewarding lifestyles as guides.

It is not always natural attributes that governments seek to preserve. The built heritage can be an important feature of a city. We have only to count the tourists who are drawn to Rome, Athens, London and Paris, and to countries such as Egypt, Turkey, China and Morocco (to present an eclectic list of heritage destinations) to appreciate the foreign earnings that ancient and/or architecturally fascinating monuments and buildings bring. A few developing countries protect and promote their unique built environment, for example Thailand with its Golden Buddha, Indonesia with the Borobudur temple and China with its Great Wall.

Preserving the built environment is a challenge if the owner of a building having heritage value is keen to demolish it and build a high-income earning housing apartment block or a commercial centre. We struggle with this concept in the first world, although those who look after the great old cities of the world are generally powerful enough to say *no* and take the glint out of a would-be developer's eye. Denying development seems to be much more difficult in the recently developed cities in countries such as Australia.

However, it is not always governments or town planning and heritage bodies that influence results. In the face of overdevelopment in Sydney in the 1970s, trade union workers from the Builders Labourers Federation, together with local residents groups, stood up to developers keen to demolish buildings of heritage value. The concept of banning work on building sites considered too precious for redevelopment became known as green bans. From small things, big things grow – the mechanisms developed by the Union have evolved to the point where it is expected that environmental considerations, including offsets, are mainstream elements in urban planning. However, not all countries have the luxury of a democracy that enables citizens (whether or not assisted by trade unions) to out-lobby developers. This situation is unlikely in a very poor country, where buildings of considerable heritage value are likely to be threatened by urban developers. If the would-be developers have some form of property right to the heritage structures, they will need to be compensated if they are prohibited from exercising their rights. The compensation they receive is yet another form of offsetting.

Biodiversity offsetting can be difficult

Biodiversity as a general principle of environmental management is the most recent initiative in environmental offsetting, and the most difficult due to a range of factors. Whether or not this form of offsetting can be declared a success in particular circumstances can be, and is, debated. The answer in political terms will be the public acceptance of the results of individual offsets. A failure, such as the loss of an endangered species, is likely to kill biodiversity offsetting, while a development project that has taken place while totally protecting an iconic species is likely to be celebrated. At present offsetting is viewed as an essential tool, probably the only available tool, in the tug-of-war between environmental health and what we will call old-fashioned development. There is, we suggest, a far more environmentally friendly form of development, which incorporates biodiversity offsetting.

Obviously, the opinion of both the scientific community and environmental advocates will count in judging the success or otherwise of biodiversity offsetting. Even though experience is relatively recent, there are already strong views that it is a sop to environmental advocates and the general public. Gaining governmental approval for a project on the basis of promises to offset residual negative impacts has obvious drawbacks – if the offsetting fails, it is too late. The mine, for example, is in full production. There is no turning back. The loss of an iconic animal would certainly be considered a disaster.

A public inquiry undertaken by the Australian Senate in 2014 heard several witnesses criticise the practice and principles of biodiversity offsetting. We will not outline those here, but rather independently search for both the positive and negative attributes of biodiversity offsetting.

Offsetting the residual damage of a project can be viewed as asking an awful lot. Keep in mind that it focuses on dealing with residual negative effects of a development. By definition, the residual is that which has not been able to be eliminated by conventional means; that is, it was not possible to mitigate environmental damage in the process of designing the project. Some would assert that this should rule out the project. Notwithstanding that, we need to ask, is enough thought and effort being put into mitigation? There are countless examples of potential negative impacts being eliminated at the design stage. It is possible that offsetting, particularly if it can be done by a monetary contribution and placing the problem in the government's hands, is too easy an option.

Altering the route of a road to diverge from a straight line to leave intact a desired natural area or historic building is a common example of design to avoid negative impacts. Those who travel along the Gold Coast Road discover a small kink in the road between Miami and Burleigh Heads. The kink is there to leave undisturbed a bora ring once used by the local tribal people.

Mitigation through design generally costs more than the original design. And the less costly a project, the better from a developer's perspective. The challenge for a developer faced with a residual impact should be to compare the cost of offsetting to the cost of mitigation through design. It is the responsibility of those who are tasked with authorising projects to make this trade-off clear to developers.

Do we acknowledge that there could be (will be?) occasions when we are going to give up on our search for a like-for-like offset? The fact that in Australia offsetting laws allow for monetary compensation suggests that those who framed the laws considered the likelihood of an unsuccessful search. This type of monetary offsetting plays an extremely important role. It can allow a project to proceed regardless of the environmental damage it causes. In theory, the amount of money paid into an offsetting fund should be enough to, for example, purchase an area of natural land as valuable in biodiversity terms as the land which is lost. This approach is useful as far as it goes – but it can be far short of like-for-like. Clearly, monetary payments will not bring back an extinct species. Monetary offsetting can be dangerous.

The nuts and bolts of biodiversity offsetting

Tor Hundloe and Shelley Burgin

In this chapter, we present a brief overview of the offsetting framework in use in Australia. As noted, Australia is a leader – in some cases *the* leader – in the assessment of the utility or otherwise of biodiversity offsetting. Given the agreement by the Commonwealth government and the state and territory governments not to have duplicate processes in the evaluation of projects that fall under both levels of government, it is possible to describe a generic approach to biodiversity offsetting. But we must note that notwithstanding this agreement, there are specific environmental laws at each level of government that, in certain circumstances, override a common approach.

Endangered species in offsetting

Endangered or otherwise threatened species require special treatment in biodiversity offsetting. There is not necessarily agreement between the national government and the state/territory ones in determining the status of animals and plants under threat of extinction, and consequently the nomenclature is different. A species deemed endangered by one level of government might have a lower threat assigned to it by the other level of government. This is important if there is a risk of extinction, hence determination of the level of that risk is crucial.

The IUCN (International Union for Conservation of Nature), of which the Australian government is a member, has established and documented in its Issues Brief (Monty *et al.* 2017) a sensible policy on where to draw the line on offsetting: 'Biodiversity offsets must not be used when a project may result in the extinction of a species'. This implies that projects or developments should not be considered if extinction is a genuine possibility; that is, a determination should be made upfront, before a developer and a government engage in the assessment process, of the threat of extinction. If it is too high, the project is automatically ruled out. The principle is clear, the practice is nowhere near clear.

No net loss

If there is one overarching philosophical principle of biodiversity offsetting it is *no net loss*. The Australian government, in its publication *How to Use the Offsets Assessments Guide*, uses the following basic equation:

$$\text{Impact} + \text{Offset} = \text{Maintenance of the Ecosystem or Improvement}$$

Maintenance equates to no net loss. The 'or improvement' is undoubtedly positive. However, it must be recognised as wished for rather than expected. There is evidence which suggests it has not been achieved in any biodiversity offset project.

Like-for-like

This other fundamental principle is stated thus: 'The quality of the offset site must at least be equal to the impact site.' There can be no quibble with this. The difficulty is locating an equivalent site. We can imagine two neighbouring blocks of, say, rainforest. One is privately owned and the owner has the right to treat it as they wish, even if they want to flatten it; however, it remains in pristine condition. In this hypothetical case, the second block is to be cleared of trees to construct a runway for the operation of aircraft in emergencies. Construction of the airfield would be allowed if the first block was purchased and brought under the control of an authority responsible for protected areas, and say, designated as a national park or equivalent status. This is like-for-like and no net loss.

This hypothetical situation is highly unlikely to exist in the real world. As a consequence, those responsible for assessing offset proposals are likely to be required to make difficult decisions. Is it a like-for-like offset if the same size block of rainforest land, with as nearly as possible the same vegetation types and fauna assemblage, is 200 km distant from the block to be destroyed? What if the substitute site is 2000 km away? What if the substitute site is twice the size of the site to be developed and has the same native flora and fauna, but is in a somewhat degraded condition due to feral pigs doing considerable damage? If the feral animals were eliminated and the site regenerated, would this satisfy the like-for-like principle? What if animals or plants had to be translocated to a substitute site? So many real-world possibilities!

Government agencies and independent researchers have developed formulae which aim to bring an element of equivalence to initially unlike offsets. While the motivation and effort of those engaged in this work can be applauded, the results will remain open to challenge for the simple reason that there are no agreed commensurate measures of biodiversity. Biodiversity does not convert into dollar terms as do the variety of goods that comprise our weekly shopping list.

As a consequence of the very real difficulties involved in locating a like-for-like environment, governments have relieved the burden on developers and instituted a system of monetary payments as an alternative. This does not overcome the difficulties of locating an on-the-ground substitute area to compensate for the one that is to be degraded. Again, government authorities have developed formulae to determine the sum of money to be paid in each case. At its simplest, that sum would equate to the cost, on the open market, of a genuine like-for-like area. It is a little more difficult to arrive at a sum of money which would allow the purchase of a partially degraded block, plus enough to restore it. There is an argument that a third option is available, which is to obtain sufficient money to purchase a completely degraded block and restore and revegetate it so that it eventually is a replica of the block to be lost to development. Putting aside the practicality of the restoration, the time delay involved before the area reaches maturity should rule this option out, unless the loss of biodiversity for decades is acceptable. We appreciate that there can be arguments made to that effect. If any such arguments are to be entertained, they need to be aired publicly and consequently be subject to informed debate.

Here is an appropriate place to recognise the sterling work undertaken by Martine Maron and her associates at the University of Queensland. In recognition of the difficulties and possible dead-ends of satisfying an exact like-for-like rule, they have developed a methodology which allows offsets to be made if they meet criteria which overall result in a like-for-like outcome. An important feature of this approach is, to use their words, 'discount' offsets which take a long time to deliver an equivalent biodiversity status to the one that is forsaken for development.

While this theoretical advancement is welcome, the methodology requires further refinement. There are two important matters to be dealt with. One is that the proposed method requires, in two of the multiple steps involved, qualitative scaling (from one to 10). Obviously, this scaling is to be, or should be, based on expert advice, for example by ecologists, botanists, biologists or zoologists with relevant specialist expertise. This advice will not be available on all occasions when it is required. The weighting may be left to people who, while well-qualified environmental scientists, may not be experts in the particular field; that is, they are not experts on the feeding habits of the animal in question. As the numbers assigned in the scales can make a significant difference to an overall result, this is a matter of concern. Qualitative scaling is not precision science, but probably the best we can do at present.

People working in the biodiversity offsets area are not the only professionals to undertake qualitative scaling. In the early days of environmental impact assessments, scales of magnitude and importance, ranked from one to 10, were used for the impacts being considered. But when we witnessed a magnitude number being multiplied by an importance number we became aware that science had given way to nonsense, maybe because decision-makers are only comfortable with numbers – any numbers! This methodology lost scientific credibility after a few years. Numbers can be seductive! Obviously, biodiversity offsetting methodology must not be taken down that path.

Another issue with the methodology is a theoretical one and cannot be resolved in its application without revision; in particular, without including another element. The methodology does not address economic consequences in the future, and these can be the fundamental issue from a human perspective in a like-for-like framework. As an example, assume that a mangrove forest, the only one in an area, supports a commercial fishery. It is bulldozed to build an airport. Adjacent coastal land of similar size would, if covered with a mature mangrove forest, support the fishery. This land is acquired from the farmer who owns it and mangroves are planted. However, from the date the original mangroves are destroyed until the new forest matures, few fish exist in the adjacent waters. It is not the abundant fishery of the past. Over time, with the developing maturity of the mangrove ecosystem, more and more fish and other animals are able to live in the surrounding waters. But by the time it is feasible to re-establish a commercial fishery, fish eaters have become consumers of similar but imported products. In this situation, the offset has succeeded but a like-for-like human environment has not survived. Maybe this does not matter. In practical political terms, we are not that sure. The fishers have a say. It is understandable that the team who developed the methodology do not address the economic matters, as they are not economists. The issue of time-lags is further discussed below.

The final word here on like-for-like offsetting is that measurement is not a problem in the other types of environmental offsetting. In carbon offsetting, the common unit of measurement is carbon dioxide. The other greenhouse gases, such as methane, are readily converted to this gas. Build a solar farm as a preventive offset and it is easy to calculate the amount of carbon dioxide not emitted.

With regard to compensating employees and business who are excluded from a natural environment so that it can acquire the status of a national park, monetary compensation is usually, but not always, all that is needed by a business. It tends to be more difficult to compensate workers as like-for-like jobs are not necessarily available locally and, consequently, retraining or other forms of recompense are required. This is easier said than done. Given the right soils and weather conditions we can regrow trees, but humans have minds of their own!

Additionality

It is a requirement that the offsetting effort is something over and above what was going to happen anyway as a business decision or as required by a law unrelated to the development. That is, the offset must be additional to what was planned to happen in normal circumstances. Meeting this criterion can be elusive. How are we to know that what is being touted as an offset was not intended regardless? There is evidence from around the world that this is a serious matter (Gibbons *et al.* 2016; Maron *et al.* 2016). We do not know, and have no means of discovering, what was in the mind of a person or the managers of a company. The best we can do is undertake economic analysis to build as realistic as possible a model of what we would expect a rational person to do if offsetting was not required. We can find no evidence of this being attempted in a systematic way. Business operators would not usually willingly open their books to allow the forensic analysis required.

Transparency

Transparency means, or should mean, that any interested party is permitted to view the result of the offsetting process. If this is not allowed, the public cannot have confidence in offsetting, and that will ultimately lead to its abandonment. The public will not necessarily have the background information or the scientific expertise to gauge the success of the offset without viewing the results first-hand: the birds live in their new environment in the same numbers as seen in the old environment.

While public scrutiny is essential, scientists have a critical role in ensuring those who are doing the offsetting are meeting the established standards. Key parameters have to be measured. For example, 'x' hectares of land of a certain landscape quality are going to be degraded by the project. The solution might be to purchase exactly 'x' hectares of equivalent quality land from a neighbour and maintain it in its natural state. Both the amount of land (very easy to measure) plus the quality of the land (difficult to measure) have to be measurable. Deciding which characteristics of the land have to be measured depends on precisely what is being offset. Biodiversity offsetting is not as simple as offsetting farming land, where soil types are likely to be the crucial characteristic and anything else of lesser significance.

If translocation of animals or plants is involved, we need to assess the population size and health of the species that is translocated. These parameters need to be the same both before and after the offset. Finally, the offset provisions have to be monitored and enforced. With regard to relocating animals, evidence of past success and failures provide a useful guide as to what to expect.

Avoiding time lags

Several other issues can frustrate the effort to offset. In many cases where the task is to offset habitat for an endangered animal, a search will be unable to discover an existing substitute ecosystem; or if it does exist, it is in the wrong place to produce the ecosystem services for the animals about to lose their habitat. Or perhaps the owner of a suitable site will not sell it at a price that the developer can afford.

There is a potential alternative solution, which involves repairing an appropriately located ecosystem similar to the one that is to be lost due to development. While the field of restoration ecology has progressed in recent years, it cannot achieve the impossible. As previously noted, complex ecosystems will take decades and in some cases hundreds of years to reach a steady state (equilibrium or climax community). Even a much simpler task, such as introducing mangroves into a littoral zone, will mean decades before the trees are fully grown and providing the same habitat and nutrition for juvenile fish, crustaceans and molluscs as the mangrove trees that were removed.

These sort of time lags we can imagine. We can also imagine the plight of a population of animals being translocated from their original habitat (with adequate carrying capacity) to one not yet capable of meeting their needs. In this case the translocated population will almost certainly decline and individual animals will suffer malnutrition and illnesses due to intra-species competition. As we explain next, governments are aware of this potential problem. To quote from the Commonwealth government guidelines:

> Where there is a large time lag between an impact occurring and an offset delivering corresponding gain, there is a risk that a threatened species or ecological community will be completely lost in the wild (Commonwealth of Australia 2012).

What is the solution? In recognition of this possible dramatic outcome, government officials have formulated the concept of advanced offsets. These are described by the Commonwealth government as follows:

> Advanced environmental offsets are a supply of offsets for future use, transfer or sale by offset proponents or providers … [they] are implemented prior to the impact occurring (Commonwealth of Australia 2017).

In practical terms, what we are discussing here is sound, empirically based land use planning, at its best. We suggest how the advanced environmental offsets could work. Geologists discover and report to the departments responsible for mining and for the environment, where they have discovered a significant ore body. The government sends environmental scientists to study the area. Assume they find an endangered species. A mining plan is overlaid with a map of the habitat of the species. The environmental scientists conclude that a significant number of the endangered species will lose their habitat and need to be translocated.

Zoologists and botanists are sent to ascertain if adjacent land offers sufficient suitable habitat for the species to be translocated. This land would be purchased (by whom is an interesting question which won't delay us here) and held for when mining commences and the endangered species is relocated on it.

While this concept of land use planning is in theory not different from governments acquiring, far in advance of need, sites for future schools, hospitals, roadways, railway lines and quarries, it has not been used by governments as a tool to allow for offsetting loss of valuable natural land. The closest we get to this concept in practice is the prohibition of development on prime quality agricultural land.

The principle of advanced offsets is being promoted, yet it is not happening in practice. Only government officials have the knowledge of future needs and are able to apply the concept of advanced offsets. We cannot count on the private sector to play a major role, if it plays a role at all. This means that at present those who have to deal with the problem of time lags (government bodies with the power to accept or reject development projects) need a means of determining whether an offsetting lag time effectively prohibits the development. The government bodies responsible for environmental offsetting have developed formulae that they believe are robust enough for the purpose of making this determination.

The process includes assigning weights (based on quasi-scientific risk assessments) to the range of natural conditions and variations to them, which in aggregate (key variables are multiplied) lead to an overall score to determine the efficacy of a potential offset proposal. Sets of equations suggest both mathematical precision and scientific rigour where, in reality, one or the other, or both, can be lacking. While the effort is admirable, the outcome is likely to be questionable.

Conclusion

We have raised issues, some of which can be resolved relatively easily but others which will challenge us and maybe remain unresolved. Then what? The one thing that is most likely is that environmental offsetting, particularly offsetting biodiversity losses, will have changed, possibly significantly, the next time it is scrutinised.

4

Management of protected areas: the need for environmental offsets

Shelley Burgin

A very strong case can be made for biodiversity offsetting on the basis that the formal protected area estate of a nation is too small and too poorly resourced to adequately deliver the
level of biodiversity a healthy, sustainable world needs. To illustrate this proposition, it is
necessary to delve into the management of national parks and similar statutory reserves. If
the extent of biodiversity protection is below the necessary level, this is the result of insufficient funding, whether provided by nature-loving tourists or allocation from annual government budgets.

The establishment of protected areas

Globally, lands have been protected since antiquity (Margules and Pressey 2000; Phillips
2004; Dudley *et al.* 2005). While historically such protection was afforded for a range of
reasons, typically it was for 'royal hunting' (Menzies 1992; Margules and Pressey 2000),
religion and/or culture (Dudley *et al.* 2005). The concept of protecting lands explicitly for
nature conservation, their scenic values (Sellars 1997; Zahniser 1992; Margules and Pressey 2000) and ecosystem services (Margules and Pressey 2000) has tended to be a more
recent phenomenon, founded in the 19th century (Sellars 1997).

When the world's first national parks were gazetted – Yellowstone in the USA in 1872,
and Royal National Park in Australia in 1879 – the concept of protecting lands for nature
conservation was novel.

Nature conservation was not of high priority in the designation of national parks in the
early years. The gazettal of Royal National Park was not based on the current concepts of
nature conservation. In its development, mangroves and mudflats were removed and
replaced by grassed parklands. Native trees were extensively logged for road works and
exotic landscaped gardens were developed, including plantings of ~4000 exotic trees.
Areas were set aside for the acclimatisation of exotic animals, with deer, rabbits and foxes
released and encouraged to become established. The national park was also used for mili-
tary exercises. However, such interferences with nature have not been acceptable in
national parks for a considerable time.

Since the declaration of the first national parks, the transition from reserving areas
primarily for recreation to protecting them for nature conservation has been gradual
(Margules and Pressey 2000; Dudley *et al.* 2005). Indeed, it was more than half a century

Table 4.1. Criteria for the classification for protected natural and cultural areas

Number	Description of protected area
I	Protected Anthropological Areas (Natural Biotic Areas, Cultivated Landscapes, Sites of Special Interest)
II	Protected Historical or Archaeological Areas (Archaeological Sites, Historical Sites)
III	Protected Natural Areas (Strict Natural Areas, Managed Natural Areas, Wilderness Areas)
IV	Multiple Use Areas
V	National Parks
VI	Related Protected Areas (Provincial Parks, Strict Nature Reserves, Managed Nature Reserves, National Forests and Related Multiple Use Reserves, Anthropological, Archaeological or Historical Reserves)

Source: Material drawn from Dasmann (1974).

before the first major landmark in the development of an international nomenclature for protected areas. Previously, nations had independently developed criteria and, although international conventions existed, each country had its unique nomenclature and management regimes (Dudley 2008).

The first proposal to develop an internationally accepted protected area terminology occurred at the International Conference for the Protection of Flora and Fauna in 1933. Four categories of protected area were proposed – national park, strict nature reserve, fauna and flora reserve, and reserve with prohibition for hunting and collecting (Dudley 2008). The Western Hemisphere Convention on Nature Protection and Wildlife Preservation proposed a modification of the suggested international nomenclature: national park, national reserve, nature monument, and strict wilderness reserve (Holdgate 1999). A World List of National Parks and Equivalent Reserves was presented at the first World Conference on National Parks in 1962, on behalf of the newly formed Commission of National Parks and Protected Areas (Brockman 1962). It was not, however, until the second World Parks Congress in 1972 that the development of an international nomenclature with agreed definitions for each category was discussed in an international forum (Elliott 1974). In the proceedings of that conference, six protected area categories were proposed, as shown in Table 4.1.

Except for one of these criteria that was based on objectives for management, in 1978 the IUCN accepted these, and developed a further five categories. However, it was soon recognised that there were inherent issues with the criteria. The problems included the lack of clear definition of a national park, the scope covered by the system, clarity of distinction between some categories, and the fact the language was terrestrial-focused (Phillips 2004). A revised system, based on categories I–V and a modification of category VIII, was used as the basis for the development of the nomenclature that was adopted by the IUCN General Assembly Meeting in Buenos Aires in 1994 (Dudley and Parrish 2006). Accompanying the criteria for six categories of protected area (see Table 4.2), this meeting also developed the first international definition of a protected area:

An area of land and/or sea especially dedicated to the protection and maintenance of biological diversity, and of natural and associated cultural resources, and managed through legal or other effective means (IUCN 1994).

Table 4.2.   The six IUCN categories of protected areas

Category	Type	Characteristics
Ia	Strict Nature Reserve	Biodiversity protection and/or geomorphological features. Emphasis on conservation values with human access/use strictly controlled and limited.
Ib	Wilderness Area	Typically largely unmodified areas that retain their natural character/influence without permanent or significant human habitation. Protected and managed to preserve their natural condition.
II	National Park	Protection of large-scale ecological processes, species and ecosystem characteristics of area. Also provides foundation for environmentally/culturally compatible spiritual, scientific, educational, recreational and visitor opportunities.
III	Natural Monument or Feature	Protection of specific natural monument including landform, sea mount, submarine cavern, geological feature (e.g. cave) or living feature (e.g. ancient grove).
IV	Habitat/Species Management Area	Protection of particular priority species or habitat of which many may need regular active intervention to address target species/habitat requirements.
V	Protected Landscape	Areas where interaction of people and nature has produced an area of distinct character with significant ecological, biological, cultural and scenic value, and where safeguarding the integrity of this interaction is vital to protecting/sustaining the area and associated values.
VI	Protected areas with sustainable use of natural resources	Conservation of ecosystems/habitats, together with associated cultural values and traditional natural resource management systems. Generally large and mostly in natural condition with proportion under sustainable natural resource management with low-level non-industrial use of natural resources compatible with nature conservation.

Source: Burgin and Zama (2014). Owned by authors, published by EDP Sciences, 2014.

The key principles of the 1994 guidelines were that:

1) protected areas were categorised primarily on management objectives;
2) assignment to a category did not reflect the level of perceived management effectiveness;
3) while categories were internationally designated, they could be varied nationally;
4) all categories were considered important;
5) there was a gradation in human intervention implied (Dudley 2008).

The use and performance of these guidelines were subsequently reviewed (Bishop *et al.* 2004). This led to calls for a greater governance dimension for IUCN protected areas at the World Congresses in 2003 and 2004. Ultimately, the World Commission on Protected Areas Categories Task Force was formed, and a new set of guidelines developed (Dudley 2008). After extensive consultation, a modified definition of protected areas was developed:

A clearly defined geographical space, recognised, dedicated and managed, through legal or other effective means, to achieve the long-term conservation of nature with associated ecosystem services and cultural values (Dudley 2008).

The switch from gazettal of national parks primarily for recreation to an emphasis on nature conservation had occurred (Dudley and Parrish 2006; Dudley 2008). Despite the initial issues, Dudley (2008) suggested that protected areas as they were then constituted were the best approach to protecting biodiversity, and that they were also progressively being recognised as vital ecosystem services in terms of genetic warehouses. More recently, Watson *et al.* (2014) suggested that, although initially perceived to protect iconic wildlife and landscapes, expectations have evolved to 'achieve an increasingly diverse set of conservation, social and economic objectives'. The IUCN (2017) has formally recognised that protected areas offer 'vital solutions to some of the most pressing global challenges, including climate change, species extinction and poverty'.

In association with the recognition of these integrated but diverse roles, in 2012 the IUCN resolved to develop an additional objective criterion for a Global Standard – the Green List of Protected and Conservation Areas. The objective of this program was to 'encourage, achieve and promote effective, equitable and successful protected and conservation areas'. To achieve the standard of acceptance into this category of protected area there need to be sound design and planning, conservation outcomes, good governance and effective management (IUCN 2017).

Gazettal of protected areas: the status quo

In parallel with the changes in the perceived role of protected areas, there has been a rapid increase, effectively worldwide, in the gazettal of protected areas (Naughton-Treves *et al.* 2005; Burgin and Zama 2014; Watson *et al.* 2014). As indicated above, effectively all countries have now adopted protected area legislation, and have reserved areas under such legislation (Phillips 2004). Many of the associated classifications fall within the contemporary protected area categories of the IUCN.

This protected area network occupied approximately 13% of the globe's total land mass by 2010 (Jenkins and Joppa 2009). Since then, the area under protection has continued to increase worldwide (Naughton-Treves *et al.* 2005; Burgin and Zama 2014; Watson *et al.* 2014). At the time of writing, worldwide there were almost 202 500 protected areas covering 20 million square kilometres, which amounts to 14.7% of the world's terrestrial lands (including inland waters but excluding Antarctica). This coverage is only 3% lower than the 2020 target set by the UN Convention on Biological Diversity (UNEP–WCMC and IUCN 2016).

The area of ocean surface within national parks jurisdiction has increased even more dramatically than the area of terrestrial lands under such protection. For example, Woods *et al.* (2008) suggested that 0.7% of the ocean surface was protected. The UN Development Programme World Conservation Monitoring Centre and IUCN (UNEP–WCMC and IUCN 2016) reported that since 2006 the increase has been more than 4 million square kilometres. There are now 15 million square kilometres, 4% of the surface of the oceans, under protection, and the area continues to increase (UNEP–WCMC and IUCN 2018).

Despite the rapid expansion of protection of the Earth's surface, crucial biodiversity areas representing both species and habitats have not yet been incorporated into the protected area network. Indeed, less than 20% of the world's key biodiversity areas are effectively incorporated in protected areas.

However, the expansion in area and number of protected areas continues in line with the agreed target of 17% of the Earth's land surface being under protection by 2020, a target considered to be achievable. For example, between April and December 2016, there was an increase of 161 646 km^2 of terrestrial areas (including inland waters) and 3 623 637 km^2 of marine areas within national jurisdictions placed under protection (UNEP–WCMC and IUCN 2016) and, as indicated above, there is a continued increase in areas protected (UNEP–WCMC and IUCN 2018).

Despite such substantial gains in the coverage of protected areas and the associated conservation actions, biodiversity continues to decline. There are numerous reasons for this situation. A significant one is that while more land is being incorporated into protected lands, areas that are not protected but that support native ecosystems continue to be cleared. This is despite an estimated need to conserve, for biodiversity purposes, 75% of original vegetated areas with forest cover. By 2015, the area had declined to 62% (Hill *et al.* 2015).

In parallel, with the continued decline in natural ecosystems, there has been an upward trajectory in the world's population. In mid-2017, the Earth's population was 7.6 billion people, with an annual increase of ~83 million (UN 2017). This rapid increase in population results in higher demand for resources. In addition, the impacts of climate change may potentially culminate in ongoing system impoverishment, even within protected areas (Carey *et al.* 2000; Hansen and DeFries 2007; McGeoch *et al.* 2011).

Australia's protected area network

Australia is one of the 13 most megadiverse countries in the world (Brooks *et al.* 2006). A principal approach to preserving this natural heritage is to declare national parks and other protected reserves. There are currently over 12 000 terrestrial protected areas, covering 151.8 million hectares, in Australia (see Fig. 4.1). This represents more than 19.7% of the

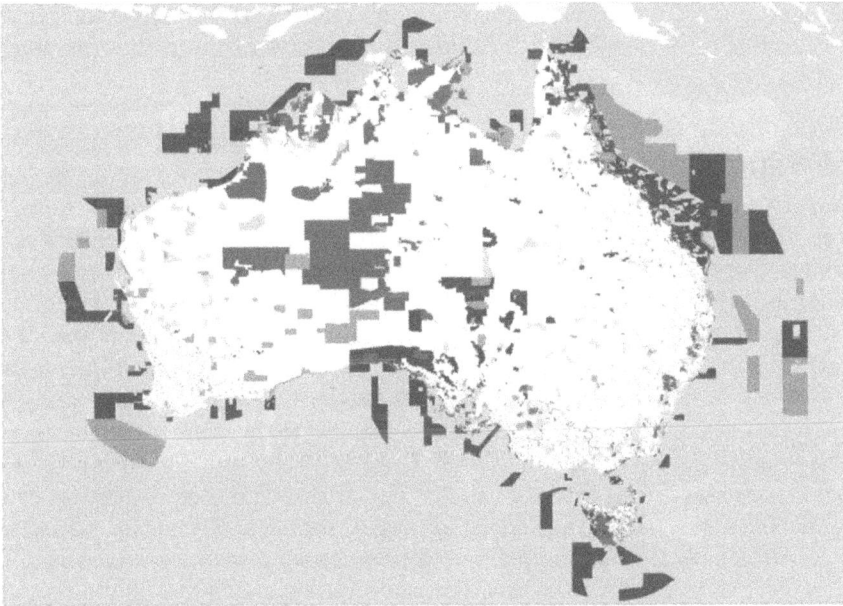

Fig. 4.1. Australia's terrestrial and marine national parks (see Clark and Johnston 2016 for colour version with details of protection categories). Source: Creative Commons Attribution Licence 3.0, Australia.

country's lands (DAWE undated). The protected areas include ecosystems that span rainforests, deserts, coral reefs and the dunes of Lake Mungo. There are 60 marine parks, incorporating some 3.3 million square kilometres. As such, 36% of the marine waters under Commonwealth jurisdiction are designated as marine national parks (Parks Australia 2020). Australia has the largest network of marine protected areas in the world (MRC undated).

In addition to having substantial coverage of lands and oceans in protected areas, Australia is at the forefront of new initiatives within this realm. For example, in 2014, three sites (Montague Island Nature Reserve, Arakwal National Park and Cape Byron State Conservation Area) were declared as pilot protected areas for the newly developed IUCN Green List of protected areas. To remain on the list beyond the pilot phase, the initial program had to be updated to demonstrate that the required performance criteria (good governance, sound design and planning, effective management) had been successfully met and that there was an assurance that performance criteria would be developed for the next phase of the program (IUCN 2017).

Expansion of protected areas

In parallel with the increasing worldwide coverage of protected areas, human populations and associated resource demands, there are indications that there is declining visitation to protected areas, or at least that visitation is not increasing despite population increases (Balmford *et al.* 2009; Buckley 2009; Hardiman and Burgin 2013).

A lessening of visitation, in particular commercial ecotourism, is a potential issue for the ongoing management of protected areas. As Buckley (2009) has suggested, visitors provide funds through entry fees and put money into local economies, that can help pay management costs. Visitors are the source of what some call political capital. Local and national politicians note the popularity of the protected areas and become more willing to allocate money for the maintenance of those areas. While it is incongruous that nature conservation, and thus funding for protected areas, rely on visitor numbers, the reality is that funding for conservation is a political decision and thus parks rely on their popularity for funding (Buckley 2009).

The current dilemma

The management of areas protected principally for nature conservation (national parks, wilderness reserves, marine reserves) varies with the jurisdiction responsible for their management. For example, in Australia, World Heritage areas are typically managed by the Commonwealth, generally in association with the appropriate state government. In contrast, national parks and many nature reserves are the responsibility of the individual states and territories. The approach to management can differ by jurisdiction. An example is the difference in national park entry charges between New South Wales and Queensland. In New South Wales, motor vehicles are charged to enter 45 of the state's national parks; the other parks have free entry (NSW Govt 2017). By contrast, within Queensland there is no charge for casual entry to national parks, although there is a fee for overnight camping (DNPSR 2017). Entry fees are an important source of revenue for management.

Despite differences in charging practices among states, and differences within a jurisdiction, protected area managers have the same duties and responsibilities. Common responsibilities include protection of the natural and cultural values, including removal of feral plants and animals, fire management and delivering nature interpretation services for visitors (Archer and Wearing 2002; Eagles and McCool 2002). For example, in Queensland the Department of Environment and Heritage Protection is charged with managing 'the health of the environment to protect Queensland's unique ecosystems, including its landscapes and waterways, … and biodiversity' (DEHP 2016).

In the absence of entry charges, funds are dependent on governments' annual budgets. As the area and number of protected areas expand, the cost of their management may be assumed to be increasing commensurately.

During the 2015–16 financial year, an additional 421 500 hectares (7.92% of Queensland) were added to Queensland's protected area estate. This increased the total land coverage to more than 4 million hectares. The state intends to add other high-value conservation areas, and lands considered of high climate change resilience. A process was also underway to nominate Cape York Peninsula as a World Heritage area, and to extend the Fraser Island World Heritage area. Apart from seeking to further protect lands within the Queensland protected area estate, support was pledged for the management of individual species including koalas and flying-foxes. All these very worthwhile initiatives cost money – and even prior to their development, the claim on the Queensland budget was considerable. For example, the 2015–16 Queensland budget allowed $148.124 million for management of protected areas. Revenue from user charges and fees (including environmental licensing for commercial activities) amounted to $50 898 million in that year; that is, of the total funding, approximately one-quarter was intended to be derived from user charges and fees (DEHP 2016).

Achievements tend to be restricted to planning, development and policies associated with encouraging visitors. For example, in the 2013–14 Queensland budget, a substantial segment of the on-ground works included infrastructure (visitor centres, skywalks, campgrounds), funding for Nature Play Queensland, and 'reduction in red tape' for 'customers' and commercial operators. Apart from fire management and response (mostly for the protection of local human communities, not primarily for management of the environment), nature conservation management was restricted to pest control and the construction of a long-range patrol vessel for the Great Barrier Reef World Heritage area. Both those initiatives were jointly funded by the Commonwealth and state government (DNPRSR 2014).

A driver of the tourism focus in Queensland's protected areas is undoubtedly to help fund management of the areas. This emphasis on tourism over conservation may result in the 'depressing slide' into environmental decline that Dudley (2008) lamented may happen. The most iconic and best-known World Heritage area in Australia, the Great Barrier Reef, was threatened several years ago with a downgrade by UNESCO to Endangered (Atfield 2013), although ultimately this did not happen. The four main factors identified as causes of degradation of the Great Barrier Reef were developments along the adjacent coast, runoff from agriculture and cities, climate change and 'direct use' (human impacts on the water). The threats were described as 'significant to the ecosystem's future functioning and resilience' (GBRMPA 2014).

It is not only Queensland that has issues with gradual degradation of protected areas. For example, the author of this chapter has been involved in a range of studies focused in the Greater Blue Mountains World Heritage area. Research made it clear that this World Heritage area is suffering from a lack of funding, even for basic environmental monitoring, despite the Blue Mountains having been a popular tourist destination since the 1860s (Burgin and Hardiman 2014).

Clearly, tourism income is not always the 'manna from heaven' that Erbes (1973) suggested it was formerly considered, even in areas where tourism has been well-developed for decades, well-advertised and destinations are perceived as 'must visit'. Without the ability to sustainably manage areas already protected for their natural and cultural heritage significance, funding issues will presumably be exacerbated as additional lands are dedicated for protection.

Environmental offsets as a source of additional funding

If the argument is made that biodiversity offsetting can make a positive contribution to the protection of biodiversity, how could this be achieved? There are two ways. One is to require a developer, who is unable to provide a like-for-like offset due to the simple fact that no land suitable as an offset is available, to pay an amount of money such that an equal or greater biodiversity gain is achieved in an unrelated but highly valuable environment. This process can substitute for the lack of funding to protect that environment. While this may be non-preferable, it could be the only pragmatic situation if the original development is allowed. The opposing view is that the development be disallowed.

Pragmatism enters into consideration if the existing protected area network is seriously underfunded and likely to remain so. The historical approach to funding the management of protected areas (from recurrent government budgets, supplemented by moneys collected as gate fees and permits) results in chronic underfunding for many protected areas. To ensure future sustainable management of a larger protected area estate, supplementary funding is required. There is no 'goose that lays a golden egg'; that is simply one of Æsop's fables (Jacobs 2001). The concept of tourism being 'manna from heaven' is probably equally fictitious (Erbes 1973). We therefore need to consider alternative approaches to funding protected areas.

The ecological-equivalent-benefit-elsewhere concept

As noted previously, biodiversity offsets were first formalised in the USA in the 1970s. Subsequently, the concept was embraced by industry, including mining, construction and forestry, to name three major ones. There were detractors, notably some environmental lobby groups. There is division within the environment movement. At one end of the spectrum, offsets are viewed as a mechanism to enhance and expand conservation initiatives; at the other extreme, they are perceived as a means to secure and maintain a licence to operate (Burgin 2008).

Despite the controversy, the concept of compensating for damage to biodiversity by offsetting with ecological-equivalent-benefit-elsewhere – a more realistic description of what is currently tagged like-for-like – has surged. Governments are increasingly recognising that biodiversity offsets can contribute to achieving conservation targets. Australia is no exception, and in fact is viewed as a world leader. However – as we identify elsewhere in this book – the lag time between loss of vegetation and replacement of habitat, if restoration or establishment from the ground up is required for a replacement ecosystem, does not necessarily constitute like-for-like and/or no net loss (Burgin 2008; Gibbons and Lindenmayer 2007; Maron et al. 2015). As discussed previously, much depends on the circumstances, such as the extent of restoration required, the time span to reach a mature ecosystem, and the success or otherwise of translocating animals. To be successful, the conservation benefits derived must equal or outweigh the loss incurred with the development (Burgin 2008; Maron et al. 2015). If the offsetting results in a gain rather than a simple replacement of biodiversity, this is undoubtedly a win.

Some advocates of biodiversity offsetting are willing to contemplate a radical departure from like-for-like, in the main because there are situations where it is not feasible. The usual alternative is for the developer to pay money to be spent on achieving a biodiversity benefit, not related to the development but nevertheless of equal or greater value than that which was lost. Maron *et al.* (2015) use the example of a developer providing funds to be spent to offset the impacts on water quality resulting from building a new port adjacent to the Great Barrier Reef Marine Park. These authors argue that, to be a valid offset, the funds should be used to enhance the water quality of the area beyond what is expected under current targets and international agreements.

Conclusion

Even in wealthy countries such as Australia, the protection of biodiversity in the government-managed protected area estate falls short of the optimum. A solution is to provide incentives to private landholders to formally agree and act to preserve biodiversity. The concept of advanced offsets was developed in relation to such incentives. For example, a farmer with an intact, forested paddock would fence it to keep domestic animals out and otherwise look after the land with the expectation that sometime in the future they would be rewarded for those efforts: a developer required to offset habitat loss would seek out the farmer and acquire rights to the protected land. The developer might then give the land to the government or otherwise have its protected area status formally recognised.

Given the very slow adoption of advanced offsets, a more realistic scenario is a developer approaching a landholder who has suitable land as an offset. Upon agreement to change the management regime (e.g. by reducing the stocking rate, fencing the creek running through the property), a financial deal would be done and a biodiversity protection outcome achieved. While the literature is scant on this type of offsetting, anecdotal evidence suggests it is the favoured procedure (other than paying money to the government) at this early stage in the history of biodiversity offsetting. It is a means of obtaining off-reserve biodiversity protection.

CASE STUDIES

5

The 'Green Olympics' saves the green and golden bell frog

Shelley Burgin

The 2000 Olympic Games were held in Sydney. They were known as the 'Green Games'.

The green and golden bell frog (*Litoria aurea*) is large (85 mm) for an Australian native frog (Cogger 2014). Historically, it was found mainly in and near coastal regions across much of eastern New South Wales from Brunswick Heads in the north to the New South Wales–Victorian border in the south (DEWHA 2008–09; OEH 2015). The former inland extremes of the species' distribution included Bathurst (White and Pyke 1996; Pyke and White 2001; Penman *et al.* 2008) and the Southern Tablelands (Osborne *et al.* 1996; Penman *et al.* 2008; OEH 2015). In Victoria, this species occurs mainly in the lowlands of eastern Gippsland (Fig. 5.1; Penman *et al.* 2008; DEWHA 2008–09; OEH 2015).

The green and golden bell frog was once common throughout its range (Goldingay 1996; White and Pyke 1996; White and Burgin 2004) and within the Sydney region it was considered one of the most common frogs (Fletcher 1889; Pyke and White 2001; Lemckert 2010). In Victoria, the species is formally considered secure (Gillespie 1996; White and Pyke 1996) although Pyke *et al.* (2002) suggest that such assessment should be treated with caution since many sightings recorded in Victoria were based on a single individual. In New South Wales, the species has been more thoroughly researched, and thus knowledge of its status, past and present, is more certain than in Victoria.

The consensus is that there has been a major decline in the green and golden bell frog in New South Wales (White and Pyke 1996; Threlfall *et al.* 2008; Stockwell *et al.* 2010). At more than one-third of the 55 known sites in the state, many within Greater Sydney, the species had become extinct or probably extinct by 2008 (DEWHA 2008–09). The remaining populations were typically small, disjunct and scattered across the frog's former range (OEH 2015). Indeed, the decline has been so dramatic in New South Wales that, by 1996, White and Pyke suggested that the species was extinct in at least 90% of its former range. By May 2016, it was estimated that only 43 key populations in New South Wales were 'known or considered likely to persist' (Dowle 2016). In preferred habitats, particularly at

Fig. 5.1.   Green and golden bell frog, *Litoria aurea*. Source: LiquidGhoul/Wikipedia, CC BY-SA 3.0.

lower elevation, this frog had previously had a continual distribution throughout the east of the state (DEWHA 2008–09; OEH 2015).

In response to the green and golden bell frog's decline within New South Wales between the 1960s and 1992 (White and Pyke 1996), it was listed as endangered under the state's *Threatened Species Conservation Act 1995*, and as vulnerable in the Commonwealth *Environment Protection and Biodiversity Conservation Act 1999* and the IUCN Red List of Threatened Species (IUCN 2016). In recognition of the species' status, recovery management plans were developed for many of the remaining populations. These plans require identification of factors that have caused the target species to decline (Mahony *et al.* 2013). This is where biodiversity offsetting has a role. Arguably, the highest-profile environmental offset undertaken in Australia for any frog species was implemented for the green and golden bell frog – for the Sydney 'Green Olympics' (Figs 5.2 and 5.3).

This chapter presents the history of the offset for the population of green and golden bell frogs in Sydney Olympic Park (the site of the 2000 Sydney Olympic Games) followed by an assessment of the outcome. As there are many basic issues, in particular, the various threats to the species that are relevant to the success or otherwise of the offset program, these are presented in an appendix at the end of this chapter.

Sydney Olympic Park: offset for the green and golden bell frog

Green and golden bell frogs were discovered in a derelict brick-pit during preparation for the development of the site for Sydney's 2000 Olympic Games (Mahony *et al.* 2013). The population was considered a significant remnant population within New South Wales (Greer 1994; Pyke and White 2001; Darcovich and O'Meara 2008). This population of frogs, within Parramatta Local Government Area, was considered one of three key populations of the species within Sydney. The other two substantial populations were in the

Fig. 5.2.   State Brickworks, 1911–12. Photo by Rex Hazelwood. Source: ON 151/24–30, State Library of New South Wales.

adjacent Local Government Areas of Holroyd and Auburn. These three, now disjunct, frog populations may be assumed to have each inhabited more extensive areas before urbanisation. The destruction of habitat associated with urbanisation would have resulted in fragmentation of habitat (DECC 2008). These three discrete populations were likely to have been part of a meta-population.

Fig. 5.3.   State Brickworks site reinvented as green and golden bell frog habitat complete with viewing platform that encircles the reconstructed wetlands. Source: Amalie Wright.

Changes to the environment for the green and golden bell frogs would have commenced when the site formed part of the large agricultural estates of Newington and Homebush. These lands were surrendered to the New South Wales government early in the 1900s and the area was subsequently used for government-owned industries, including the State Brickworks, State Abattoir and Royal Australian Navy Armament Depot (Beder 1993; Darcovich and O'Meara 2008), and for private industries including Union Carbide (Lenskyj 1998). Other industries including an oil refinery, chemical industries, a petroleum products storage area and a fuel terminal were situated on adjacent areas. There was also a landfill for domestic, commercial and industrial waste, estimated to cover 9 million cubic metres (Beder 1993).

There was extensive pollution of the landscape with dioxin, asbestos, heavy metals and phthalates (Beder 1996; Lenskyj 1998). In the 1980s, the industries on this 800 hectare site were removed (Darcovich and O'Meara 2008), leaving behind what Beder (1993) described as 'one of Australia's worst toxic waste dumps'. Despite this degradation, the site was considered of substantial ecological value, largely due to its isolation from the urban development of Sydney (Greer 1994). It was only 15 km from the city's CBD.

When quarrying of shale and sandstone ceased in 1988 and 1992, respectively, and an adjacent abattoir was demolished, freshwater wetlands developed in the disused brick-pit and its environs, and pioneering vegetation colonised the site (Darcovich and O'Meara 2008). During an ecological survey in 1993 green and golden bell frogs, including tadpoles, were observed in the brick-pit and its surrounds. The brick-pit was the core habitat of the frog population. This frog population was considered the largest and most viable within the Sydney region (Greer 1994; Pyke and White, 2001; Christy 2003).

Later in 1993, the International Olympic Committee announced that Sydney would host the 2000 Summer Olympic Games. The site was chosen. This required fast-tracking the remediation of the land, and the commencement of construction of the infrastructure required for the Games. Before that could start, however, the transition of the site to the Olympic Committee required approval under the NSW *Environmental Planning and Assessment Act 1979*.

By this time the green and golden bell frog was listed as endangered under the NSW *National Parks and Wildlife Act 1974*. A licence to 'take or kill' frogs which were encountered during development was required, as was the preparation of a Flora and Fauna Impact Statement. With the introduction of the *Threatened Species Conservation Act 1995*, development applications required an '8-Part Test' (Darcovich and O'Meara 2008). Had the frogs not been endangered the approval processes would have been satisfied with offsetting the potential loss of the frogs; because of their endangered status they were not allowed to be relocated (assuming that was feasible). They had to be saved by modelling the wetlands on the Olympic site to ensure their survival. In fact, the requirement went further than this and sought to have the wetland habitat values enhanced. This was potentially a major win for the green and golden bell frog.

The Flora and Fauna Impact Statement confirmed that the frog population was centred on the brick-pit, although individuals were resident over a much wider area of the Olympic site (Pickett *et al.* 2013).

The initial plan for the Olympic site was to remove all brick-pit frog habitat in order to build a tennis complex and an amphitheatre. If this was to be allowed, a like-for-like offset was necessary. It was proposed that four widely spaced ponds be developed to provide the necessary habitat. If that was insufficient, it was proposed that additional wetlands could be constructed. The idea was that the ponds would offer habitat to a larger population of frogs than the number inhabiting the brick-pit. These offset proposals were rejected, and the tennis courts and amphitheatre were developed elsewhere on the site (O'Meara and Darcovich 2015). If nothing else, this proves that not all offset proposals are accepted.

The subsequent plans for development of a new green and golden bell frog habitat also had to be modified. The proposed development presented in the Masterplan of Homebush Bay Corporation showed that a substantial area of ephemeral waters in the brick-pit would be impacted and thus inhibit migratory movement of the resident frogs. For this reason, that proposal was also rejected (AMBS 1999).

The development that ultimately occurred resulted in nine of the 26 existing ponds in the brick-pit being lost due to flooding of the lower levels (AMBS 1999). However, 19 ponds were to replace the nine that were lost. It was also required that an additional 24 ponds be constructed elsewhere in Sydney Olympic Park.

The success of the construction works in the brick-pit depended on these ponds being colonised by the green and golden bell frog. To ensure the ongoing viability of the population, the management of 200 hectares of frog habitat included continuing research and monitoring. Over 70 frog ponds, terrestrial habitat, frog fences and underpasses were constructed and maintained (O'Meara and Darcovich 2015).

By 2001, a year after the Games, the estimated population of the green and golden bell frogs in Sydney Olympic Park was ~1500, ~600 of them in the brick-pit. An additional 900 individuals inhabited the constructed wetlands within Olympic Park (Campbell 2001). However, there was a later downward trend in frog numbers. In response to this decline, and the suggestion that inbreeding depression may be a threat to the long-term viability of the population, a breeding and translocation program was introduced.

There appears to be only limited scientific support for the concern about inbreeding (Colgan 1996; Greer 1996; Burns et al. 2004). Burns et al. (2004) found no evidence of a genetic bottleneck at the population level – indicating that inbreeding depression was not an issue. In response to threatened stochastic extinction, however, Pickett et al. (2013, 2014) recommended artificial reintroductions of frogs and improved connectivity among precincts within Sydney Olympic Park. White and Pyke (2008a), who were involved in the reintroduction project, considered that this should remain a last resort.

Initially there was a significantly increased population (Campbell 2001). This was followed by the establishment of a sustained population level due to recruitment (SOPA 2007, 2009, 2011, 2012). The more recent decline in numbers (SOPA 2015) has not been great enough to warrant going to the 'last resort'.

A success story in biodiversity offsetting?

The program to conserve the green and golden bell frog population in what became Sydney Olympic Park has been ongoing since 1993, when the species was discovered onsite. The outcome has been maintenance of the core population within the brick-pit, together with the development of two sub-populations in constructed habitat on remediated and/or restored lands (Narawang wetland, Kronos Hill-Wentworth Common). More recently, there has been an additional development of habitat at Blaxland Riverside Park (SOPA 2015).

The focus of the offset was to enhance and maintain onsite a viable, free-ranging population of the endangered green and golden bell frog. This required management of the frog population throughout major remediation and development works. Over this period, the resident number of frogs in the brick-pit increased, and sub-populations developed. This is success, despite a downward trend in numbers and activity more recently (SOPA 2015).

The population is more secure than when first identified in the brick-pit. Indeed, the current decline may simply reflect a population adjusting to its carrying capacity.

The result: *'The winner is ...'*

The focus on 'green' in the bid for the 2000 Olympic Games (Beder 1993; Cashman and Hughes 1998; Lenskyj 1998, 2000) was a pertinent move since the International Olympics Committee modified its charter in 1996 to include reference to the environment, and in 1999 created the *Olympic Agenda 21* (Ashton 2016). Within Australia, the emphasis on 'green' was widely noted in the media. For example, Bruce Baird, the New South Wales Minister responsible for the Olympic bid, was reported to have stated 'no other event at the beginning of the 21st century will have a greater impact on protecting the environment than the 2000 Olympic Games in Sydney' (Beder 1993; Lenskyj 2000).

The Commonwealth Minister responsible for the environment, Ros Kelly, also claimed that winning the 2000 Olympics bid would be a 'vote for the environment' (Beder 1993).

In a society where the community will much more readily engage in the conservation of 'cuddly koalas and smiling dolphins' (Lenskyj 1998) than focus attention on a frog, no matter how iconic, the outcome of the offset for the green and golden bell frog in Sydney Olympic Park has to be viewed as a major success. However, of great significance in this success was the discovery of a species that as recently as 1991 had been declared endangered in New South Wales (Lunney and Ayers 1993; Lunney *et al.* 1995) and found in a most unlikely place, a derelict brick-pit within a heavily degraded and polluted landscape. That the animal was emblazoned in green and gold, Australia's national sporting colours, added to the interest in the discovery. These attributes provided the perfect background for what Lenskyj (1998) described as a 'media bonanza'. It was certainly a bonanza for the resident population of frogs. The species could not have been ignored in a bid for the Green Olympics. Indeed, under the circumstances, it would have been inconceivable not to ensure the protection of the species.

Initially, management of this bell frog population was apparently extremely successful. For example, when Greer (1994) undertook the first survey of the brick-pit he estimated that 55–110 green and golden bell frogs were present. By 2001, the population was estimated to be 600 in the brick-pit and 900 in constructed habitat elsewhere in Sydney Olympic Park. In 2006, the population was reported to 'remain strong and relatively stable' (SOPA 2006). Since that time the annual reports of Sydney Olympic Park have indicated that, for much of the intervening period, the population remained effectively static (SOPA 2007, 2008, 2009, 2011, 2012); although, as already indicated, there has been a downward trend in the numbers and activity of the frog. Despite this downward trend, Sydney Olympic Park Authority (SOPA 2016) reported that the breeding population was 'relatively large' and that breeding occurred across all precincts which had the appropriate habitat.

In addition, a breeding colony has been developed at Taronga Zoo (McFadden *et al.* 2008) and there has been a reintroduction of the species to areas of its former range (Daly *et al.* 2008; Stockwell *et al.* 2015; White and Pyke 2008a). These initiatives are a likely consequence of the success at the Olympic Park site.

Community and scientific groups (Frog and Tadpole Study Group; Australian Herpetology Society; Royal Zoological Society of New South Wales) have rallied to raise the profile of the species. For example, the proceedings of the only two symposia on frogs published by the Royal Zoological Society of New South Wales focused on the green and golden bell frog (Pyke and Osborne 1996; Goldingay and Osborne 2008). All indications are that the green and golden bell frog was a winner, as Australia was in holding the first Green Olympics.

Not all are winners: *'The losers are...'*

In all the publicity associated with the offset to save the endangered frog on the Olympic Games site, it appears to have been forgotten that it is not the sole species of frog in Sydney Olympic Park. Indeed, there are seven species. While the endangered green and golden bell frog benefited directly from the initial $6.5 million that transformed the brick-pit frog habitat into an iconic venue, the other frog species in the Olympic Park have been effectively ignored.

The implementation of the offset for, and the ongoing management of, green and golden bell frog habitat has enhanced the size of the population of this endangered species, and has undoubtedly expanded the scientific and management knowledge of the species. However, it is unknown whether the substantial effort has benefited any of the other resident frog species of Sydney Olympic Park. If it has, it probably has been pure serendipity.

Beyond the confines of Sydney Olympic Park, there does not appear to have been a major shift in public engagement with frogs. For example, although the green and golden bell frog may have been one of the most common frogs in Sydney (Fletcher 1889; Lemckert 2010; Pyke and White 2001), the green tree frog (*Litoria caerulea*) was also extremely common, and should have been much more difficult to ignore than the green and golden bell frog. The sharp decline in the green tree frog occurred effectively without note, even by this writer (White and Burgin 2004).

The unnoticed decline in the green tree frog occurred despite the species living effectively in a symbiotic relationship with humans. It heralds the rain with its loud call, magnified by its echo within house downpipes. It was often obvious of a night, clinging to the outside of house windows, waiting to collect moths and other prey attracted by artificial light.

The stark contrast in levels of concern for the plight of these two species – the green and golden bell frog and the green tree frog – is evidenced by the output of the *Australian Zoologist* (the journal of the Royal Zoological Society of New South Wales) between 1992 and 2017. While the focus of 65.5% of the 87 published papers that discussed frogs (including those from two symposia specifically on bell frogs) were directly related to the green and golden bell frog, not a single paper focused on the plight of the green tree frog.

Conclusion

Without the offset, the population of green and golden bell frogs would very likely be at least functionally extinct or close to it. Instead, the current breeding population in Sydney Olympic Park is significantly larger than when first encountered in the brick-pit and its surrounds. This success demonstrates that biodiversity offsets can be a valuable tool for conservation. However, the financial costs involved have been significant. Land and real estate developers could object to spending as much money as governments did.

For an indication of the actual complexities involved in what might seem a rather simple biodiversity offset exercise, as described in this chapter, reading the Appendix is highly recommended. Biodiversity offsetting, particularly if it involves significant habitat reconstruction, requires much knowledge of the animal or plant in question. The biological complexities involved mean it is not a job for lay people. Read on!

Appendix

Key threats in the management of green and golden bell frog populations

In Sydney and beyond (Pambula, South Coast) recovery actions for the green and golden bell frog commenced in 1993 (White and Pyke 2008a). However, despite substantial research and translocation attempts, failure has been frequent (White 1998; Pyke and White 2001; Pyke *et al.* 2008; White and Pyke 2008a). The key impediments included habitat fragmentation, disturbance, modification or loss of certain habitat characteristics (Mahony 1996; Pyke and White 1996, 2001; Hamer *et al.* 2002); presence of predatory fish (Pyke and White 2001; Hamer *et al.* 2002; White and Pyke 2008a); disease (Daly *et al.* 2008; Penman *et al.* 2008; Stockwell *et al.* 2015); and proximity to a source population (White and Pyke 2008b). Each of these factors was identified as likely to have contributed to, or caused, the decline of the green and golden bell frog in the Sydney Olympic Park area (SOPA 2015).

Habitat fragmentation, disturbance, modification and/or loss of physical habitat characteristics

In most situations where offsets are required due to habitat being modified, there is a substantial lag time between the loss of habitat and the creation of replacement habitat. When a green and golden bell frog habitat area is to be redeveloped predominantly as an 'eco-tourism exhibition' (the terminology used to describe the end result of the Sydney Olympic Park redevelopment), we do not have a like-for-like offset. For example, while wetlands such as those in the Sydney Olympic Park area can be redeveloped to mimic natural habitat, the offset areas were incorporated in lawns with fences and underpasses. The new habitat was very different from the original habitat. As only one example, the movement of the frogs was restricted.

Unusually for an endangered species, the green and golden bell frog is a coloniser or, as Pyke and White (2001) suggested, a 'weedy' species. Such colonising species typically invade and become established in newly formed or disturbed habitat. This is a beneficial characteristic in the right circumstance. That circumstance is crucial. If transformation of a landscape changes the dynamics of the ecosystem to the extent that a pioneer species does not find it suitable, it will not survive there. Even if the animal does colonise the new habitat, there can be problems in the future as succession takes place. A redeveloped ecosystem is not stable for a considerable length of time. The downward trend in activity and numbers of green and golden bell frogs in Sydney Olympic Park is possibly (or likely) to be related to successional change in some attributes of the frog's habitat.

White and Pyke (2008b) compared changes in wetland habitat in the area between 1995 and 2007; that is, before and some years after the Olympic Games. During this period, the extent of open water and submerged/floating vegetation more than doubled while emergent vegetation increased by at least 80%. This, together with the observations that fringing vegetation (used as shelter and chorusing sites) and the foraging area had not changed despite frog numbers increasing, implies that competition for preferred resources increased, particularly for preferred water depth and cover – two attributes previously implicated in the lack of success of translocated green and golden bell frogs (White and Pyke 2008a).

Vegetation and pond hydrology are managed on an ongoing basis in Sydney Olympic Park (SOPA 2014). However, it has been suggested (Campbell 2001) that as the population of frogs has increased in the artificial wetlands and the vegetation has matured, other attributes of the dynamics of the frog's population may also have changed, resulting in increased competition for resources. For example, with the increased numbers of green and golden bell frogs, competition for preferred foods could result. Likewise, competition for habitat, including breeding sites, may result in greater predation and/or reduced breeding success among resident frogs.

Stockwell *et al.* (2015) suggested that competition and predation could result in decline in frog populations if the target species was prey or the species was an inferior competitor. Pyke and Miehs (2001) observed that adult green and golden bell frogs voraciously consumed emerging metamorphs of their own kind, indicating that a decline in the population may be due to cannibalism. Such intra-specific competition, at least among tadpoles, increases in intensity when food is limited (Mokany and Shine 2003). In terms of predators, fish are considered a significant predator of the frogs (Mahony 1996; Pyke and White 2000; Hamer *et al.* 2002).

Presence of predators

The green and golden bell frog is particularly vulnerable to predation by the mosquito fish or plague minnow (*Gambusia holbrooki*) (Mahony 1996; Pyke and White 2000; Hamer *et al.* 2002). Management actions to protect frogs in Sydney Olympic Park include drainage of wetlands to remove mosquito fish (O'Meara and Darcovich 2015). This approach reduces such predation to an intermittent issue, as indicated by the surveys of White and Pyke (2008b) in 1995 and 2007 which did not observe mosquito fish in any wetlands surveyed in the area.

There is no doubt that, when present, mosquito fish consume the eggs of green and golden bell frogs, together with their fry and tadpoles (Pyke and White 2000). However, in a review of the decline of this species of frog between 1990 and 1995, White and Pyke (2008b) found no association between the loss of the frog populations and the presence of mosquito fish. Co-existence of the two species is presumably due to the presence of habitat characteristics that provide protective cover for frogs, or the relative levels of fish and frog populations being insufficient to cause local extinction of the frogs.

While mosquito fish are undoubtedly a voracious predator of green and golden bell frogs, other predatory fish also consume the eggs, tadpoles and/or fry. In laboratory comparisons of predation rates on this species of frog between mosquito fish and empire gudgeon (*Hypseleotris compressa*), firetailed gudgeon (*H. galii*), Pacific blue-eye (*Pseudomugil signifer*) and red-fin perch (*Perca fluviatilis*), all species consumed the eggs, tadpoles and fry of the frogs. However, all except the mosquito fish did so only when hungry and, even then, predation by gudgeons, blue-eye and perch was limited compared to that of mosquito fish (Pyke and White 2000).

In addition to fish predators, many other species are known or likely predators of green and golden bell frogs. Such taxa include birds, snakes, turtles, aquatic invertebrates and urban predators including the black rat (*Rattus rattus*), red fox (*Vulpes Vulpes*) (Pyke and Miehs 2001), cats and dogs (Goldingay 1996). As already indicated, cannibalism also occurs (Miehs and Pyke 2001; Pyke and White, 2001; Thurley and Bell 1994). Direct evidence of predation at any stage of the green and golden bell frog lifecycle is, however, generally limited (Pyke and Miehs 2001; Pyke *et al.* 2013). One exception is cannibalism, and predation by the water skink (*Eulamprus quoyii*). Predation is likely to occur when tadpoles and metamorphs are present. Under such circumstances, adult green and golden bell frogs and water skinks were observed to congregate at the edge of ponds, and both species preyed voraciously on the young frogs. Metamorphs were particularly preyed upon by both species as they emerged from the water (Miehs and Pyke 2001; Pyke and Miehs 2001).

Disease

Chytrid fungus (*Batrachochytrium dendrobatidis*) has been responsible for the decimation and extinction of frogs worldwide (Stuart *et al.* 2004; Pounds *et al.* 2006; Skerratt *et al.* 2007). The impact on Australian frogs has resulted in the fungus being classified as a key threatening process under the New South Wales *Threatened Species Conservation Act 1995* and the Commonwealth *Environmental Protection and Biodiversity Conservation Act 1999*.

Outbreaks of chytridiomycosis caused by chytrid fungus infection have occurred within Greater Sydney (Burgin *et al.* 2005; White and Pyke 2008a) and impacted green and golden bell frogs elsewhere in their range (Daly *et al.* 2008; Penman *et al.* 2008; Stockwell *et al.* 2008). Indeed, the chytrid fungus has been proposed as the main cause of decline in these frogs throughout their New South Wales range (DECC 2005), hence the designation of the fungus as a key threatening process in the decline of this frog species.

There has been one reported outbreak of chytridiomycosis in green and golden bell frogs of Sydney Olympic Park. This was identified during routine monitoring (November–May, 1998–2005) across the 200 hectare area of Olympic Park designed as habitat for these frogs. Over 55 survey nights, a total of 23 moribund green and golden bell frogs were encountered, generally with five or fewer found diseased in any of the surveys. All deaths were associated with a small complex of ponds within a 4 hectare area. Of the 23 bell frogs, 17 were identified to have succumbed to the effects of chytrid fungus. No further outbreaks of disease were observed, and despite the spike in deaths, the range of variation in numbers of frogs encountered in subsequent surveys appeared unaffected by the observed mortality event (Penman *et al.* 2008). The implication is that the population had recovered.

Elsewhere in Sydney, Burgin *et al.* (2005) observed similar isolated outbreaks and subsequent recovery of local populations of three different native frog species. Due to the subsequent recovery of frog populations, the authors concluded that chytridiomycosis was not the proximate cause of deaths. However, they did not consider whether temperature maxima at low elevations of Greater Sydney was a factor in the survival of the fungus in some years.

Such a situation is consistent with the disease occurring intermittently. Other outbreaks of chytridiomycosis in green and golden bell frogs have occurred in rehabilitated and supplementary populations: various sites in Sydney (Pyke *et al.* 2008; White and Pyke 2008a); Pambula on the South Coast (Daly *et al.* 2008); and Hunter wetlands (Stockwell *et al.* 2008). The loss of released green and golden bell frogs is also consistent with the hypothesis that temperature may be a factor in such intermittent outbreaks. For example, White and Pyke (2008b) suggested that low water temperature was a reason for loss of translocated green and golden bell frogs. If this were the situation, it would be expected that, as Burgin *et al.* (2005) observed, the impact of chytrid fungus would be greatest in late winter. While it may be assumed that dead and dying frogs would be noticed if such an epidemic continued into late winter, surveys by Penman *et al.* (2008) did not span winter, and scavengers tend to rapidly remove dead frogs (Wotherspoon and Burgin 2011).

In Sydney Olympic Park, deaths from chytrid fungus could therefore have continued into winter. Support for the potential for the disease to be sustained into later winter without being noticed is provided by Stockwell *et al.* (2008), who warned that the absence of dead and dying frogs does not preclude the presence of the chytrid fungus. However, the observation that there was no change in the range of variation in surveys of the population of green and golden bell frogs in Sydney Olympic Park after the identified epidemic (Penman *et al.* 2008) does not preclude the loss of frogs and subsequent repopulation of the ponds by immigrants. However, this would depend on the proximity of a source population.

Proximity of source population

It has been suggested that inbreeding depression may be responsible for the recent downward trend in activity and numbers of green and golden bell frogs within Sydney Olympic Park (SOPA 2015). As outlined above, this population could be assumed to be a remnant of a much more extensive population. With isolation, the subsequent genetic bottleneck may result in the stochastic loss of heterozygosity, leading to inbreeding depression. To address this potential issue, a breeding program was introduced (SOPA 2015) despite an apparent lack of evidence that inbreeding depression was an issue (Colgan 1996; Burns *et al.* 2004). It is also likely that some immigration is involved. For example, elsewhere in Sydney, green and golden bell frogs were found in newly constructed wetlands on the Macquarie University campus (Joss pers. comm.), an area surrounded by barriers created by urban infrastructure including major highways. The outcomes of the breeding and translocation program introduced in 2013–14 (SOPA 2015) allow for assessment of such effects.

Conservation of the bell frog

A major complication with ensuring the long-term viability of green and golden bell frogs is that many remaining populations within New South Wales are in coastal/near coastal habitat (White and Pyke 1996, 2008b) – areas of high human habitation. For example, Sydney had the fastest-growing human population in Australia (ABS 2016) and between 1996 and 2007, 13 of 20 known sub-populations of the species were lost from the Sydney area. Such loss is not unique to Sydney (White and Pyke 2008b). Populations, such as that in Sydney Olympic Park, already considered significant (White and Pyke 1996; Darcovich and O'Meara 2008), will become increasingly important remnant populations if declines continue.

The development of the biodiversity offset, and continued management of green and golden bell frogs in Sydney Olympic Park, has been heralded as an example of the successful use of such offsets to sustain endangered frog species (Darcovich and O'Meara 2008; Pickett *et al.* 2013).

Recreational fishers seek offsets

Shelley Burgin and Alan Midgley

> Australian bass are an iconic Australian freshwater target species for east coast anglers. They are aggressive fighters that spend most of the time in and around structure of some description and ... taste like mud (getfishing.com.au).

There is concern that the Australian bass (*Macquaria novemaculeata*), a freshwater fish, has declined – certainly since Europeans arrived (Pollard and Growns 1993; Shaddick *et al*. 2011). As with other freshwater fish, many reasons can in part, in full or in combination explain population declines; for example, river regulation (dams, water extraction), degradation of habitat and competition for food from introduced fish.

For certain popular species, particularly in rivers or impoundments near large human population centres, recreational overfishing is a threat. If nothing else, the harvestable population is reduced and fishers need to spend more effort to catch the number of fish caught in the past. Fishers are quick to assign blame. Someone is denying them their right to catch as many fish as they wish! This attitude needs to be understood if we are to comprehend the values and psychological disposition of recreational fishers. There is a political party in Australia formed to defend the rights of fishers and hunters.

The right to fish is considered a god-given right by many anglers. One recreational fishing lobby group expresses this attitude less dramatically, without downplaying its social significance: 'the right to fish recreationally is entrenched in Australian society and cultural heritage' (Recfish West 2014).

Australian bass is most heavily fished by recreational anglers in the Sydney Basin (Henry and Lyle 2003) and is particularly targeted in the Hawkesbury–Nepean River (Midgley 2015). A reason for this is that the river flows through the northern, eastern and western edges of Sydney, Australia's most populous city, with over 5 million inhabitants. The fish is a preferred target species (Midgley 2015).

An indication that overfishing could be having a negative impact on the long-term sustainability of a fishery is a deficit of larger fish in the population (Conover and Munch 2002; Midgley 2015). To maintain long-term sustainability, larger fish (the individuals that produce the largest number of offspring) need to be conserved, particularly if the species is already over-exploited (Conover and Munch 2002). For decades, it has been acknowledged that the population of Australian bass in the Hawkesbury–Nepean River is dominated by small fish (Harris 1988; Midgley 2015). This suggests overfishing of the larger, more mature fish.

In the past, commercial fishing was likely to have been an influencing factor in the decline of Australian bass. Gray *et al.* (1990) reported that this species made up the greater proportion of the by-catch of the prawn trawling industry in the lower reaches of the river. Between March 1986 and February 1988, these researchers reported that 1851 Australian bass were caught as by-catch of this industry. However, that is the past – fishing of Australian bass in the Hawkesbury–Nepean River is now restricted to recreational fishers. The impact of the commercial harvest would have been rectified relatively quickly if there had been no fishing after the cessation of commercial fishing.

Before considering the potential role for offsets to support the species' long-term sustainability, we present a thumbnail sketch of the fish's biology.

Biology of Australian bass

The Australian bass has an east coast distribution, present in eastward-flowing rivers, tributaries and associated wetlands along the east coast from the Mary River in Queensland to Lakes Entrance in Victoria. The catchments of virtually all these rivers have been converted into farms or taken for urban development, whether for the expansion of Australia's largest city or for the extension of rural villages.

Australian bass is a carnivorous, temperate species. It grows to ~60 cm and may weigh as much as 3.8 kg (Allen *et al.* 2002). For a fish, it is long-lived with some individuals sampled in the Hawkesbury–Nepean River estimated to be 22 years of age (Harris 1985). More recently, the oldest fish sampled by Midgley (2015) in the same river was 28 years. The fish, a female, was also the longest (37.4 cm). The longest male he caught was 29.6 cm at six years of age (Midgley 2015). However, overall, older and larger fish are rare in the river system. In the absence of an alternative explanation, this indicates that the species is overfished by recreational fishers.

Australian bass use a diversity of habitats and survive in a wide range of salinities and water temperatures. For much of their life, individuals remain in freshwater habitat but migrate downstream to breed before returning upstream to the freshwater reaches of rivers and creeks. Such migration occurs between May and July (austral autumn through into winter). In the Hawkesbury–Nepean estuary, spawning occurs in waters that are approximately one-third of the salinity of seawater and within the temperature range of 11–16°C (Harris 1986).

Once fry hatch from their demersal egg (an egg that rests on or near the bottom of the river), they move to the protection of aquatic plants near the bank (the plants can be submerged, emergent or floating). Typically, in the first year of life, fry migrate upstream into the freshwater reaches (Midgley 2015). They remain in freshwater until they mature at two to four years for males, and five to six years for females (Harris 1986).

While Australian bass have the slowest growth rate of any species within its genus (Harris 1983) they do have the potential to reach 10 cm in the first year (Growns and James 2005). Their growth depends on environmental conditions. For example, a comparison of growth and size attained in riverine environments (Harris 1983) and impoundments (Wilde and Sawynok 2005) indicates that they grow faster in impoundments than in rivers and their tributaries. These differences would be associated with differences in environmental conditions within the different habitat, and/or level of competition.

Once mature, Australian bass migrate annually from their freshwater habitat to the estuary to spawn, although not all fish undertake this migration annually (Midgley 2015). While this may be a natural phenomenon, it would also be influenced by environmental conditions. For example, it may not be physically possible for the fish to migrate downstream in some years when water levels are low due to drought. As the species inhabits waters throughout a catchment, fish are potentially affected by human influences over the full length of the river, from the estuary to the headwaters of the main river and its tributaries, and the wetlands on the floodplain.

Although considered to be relatively abundant in most of the catchments where Australian bass occur naturally, as indicated above, since European arrival there has been concern for their decline (Pollard and Growns 1993). While the decline of any native species is of concern, Australian bass grow slowly, and in 1988 Harris reported that there was declining recruitment and increasing mortality. There is no reason to believe that the situation has improved dramatically since that time. Harris (1983) attributed the decline to exploitation and habitat damage.

There is a range of reasons for habitat damage. Chief among them is river regulation, which has been widespread in the Hawkesbury–Nepean River (and elsewhere within the range of the species). Regulation has included the development of dams and weirs that have restricted the migratory movement of the fish. As indicated above, there have also been major changes in land use including removal of natural vegetation and its replacement with agricultural lands and urban development (Pollard and Growns 1993; Shaddick *et al.* 2011). This is ongoing in the catchment (Burgin *et al.* 2016).

As indicated above, concern for the decline of Australian bass in the Hawkesbury–Nepean River was sufficiently great to ban commercial fishing of the species some decades ago (Gray *et al.* 2003). However, recreational fishing continues despite recognition of the decline, and indications are that the species continues to be under threat despite most of the catch being released (Midgley 2015). Indeed, it remains a prized angling species by both tournament and non-tournament recreational fishers (Dowling *et al.* 2010).

Potential for offsets to support the continued viability of Australian bass

The Australian recreational fishing body Recfish West (2014) suggested that when any decision threatens to compromise the values of a recreational fishing resource due to 'loss of availability, accessibility or quality', action should be taken to assess and then avoid, mitigate or, where required, offset the loss of the amenity. The organisation goes on to recommend that in developing offsets, there is a need to ensure that the conservation efforts will be of 'significant benefit to the recreational fishing resource and its stakeholders'. One approach that is widely used in Australia to support recreational fishing is restocking impoundments and waterways. This need not be to offset a decline in catches elsewhere, but it can be. And when this is the rationale, it is an offset. Stocking and restocking programs fall within the jurisdiction of government departments charged with the management of fisheries. In New South Wales, this is the Department of Primary Industries.

In 2017, almost 500 000 Australian bass were released into New South Wales dams to support recreational fishing. It was considered a bumper year for the Australian bass restocking program – the most released in a single year. There has been an increasingly significant Australian bass breeding and release program since the 1980s, with the primary aim being to support recreational fishing (DPI 2017). The focus of the releases is therefore not targeted at conservation of the species but rather, as Virtue (2017) reported, to stock impoundments for recreational fishing. The fishers view this as an offset – although they are unlikely to call it that – to compensate for the decline in catches due to the building of dams and the residential and industrial expansion into the catchments.

Recreational fishers who firmly believe that they have a right to fish – some believe that this so-called right extends to catching as many as they can – expect governments to take action to compensate for any decline in catches. These, plus some other recreational fishers, believe that their fishing licence fees provide money for restocking, among other management activities. Governments are likely to reinforce this view when setting licence fees.

What tends to be overlooked is that population growth, where it leads to more fishers, is one of several reasons for per person decline in catches. However, with releases, there is less recreational fishing pressure on the wild stock and, provided the genetic integrity of the species is maintained in the breeding program, it may ultimately contribute to reversal of the downward trend in wild populations. Additionally, media coverage of the releases has the potential to raise the profile of native fish within the broader community. Each of these attributes of restocking programs therefore has the potential to contribute to the conservation of the target species and potentially supports the longer-term viability of the species.

The support for the restocking of wetlands with native fish, including Australian bass, is evidenced by the popularity of the Dollar for Dollar Native Fish Stocking Program. This program is administered by the New South Wales Recreational Fishing Trust. On a competitive basis, the Trust offers matching funding to community groups, typically recreational fishers, for the release of Australian bass into eastern-flowing rivers. By benefiting anglers, the program has the potential to engage communities in responsible stocking and sustainable fishing (DPI undated). Restocking provides a surplus of fish since only a proportion of the fish will be caught. And, since most fishers who target Australian bass release their catch, there would be more Australian bass in the river systems.

Various offsetting projects have been undertaken, and some are still underway. For example, in the Hawkesbury–Nepean and Hunter Central rivers, the Tide to Table model was developed. This program was piloted in the Sydney region, which encompasses the Hawkesbury–Nepean River. The initial project was aimed at restoring fish habitat, including addressing poor water quality, within 30 sites. The program has since been extended to other areas.

These programs are not theoretically an offset for the Hawkesbury–Nepean River Australian bass fishery, although they provide benefits across New South Wales for Australian bass and, of course, the Hawksbury–Nepean River fishery benefits. For example, restoration of riverbanks with native vegetation will provide habitat for the terrestrial prey (mainly insects) of Australian bass and, over time, will increase the health and complexity of the instream environment and provide habitat for both the fish and their aquatic prey. Ultimately, therefore, the restoration effort will improve habitat for the fish. Thus, works that may be primarily undertaken to stabilise riverbanks would, ultimately, provide additional prey while the roots of trees and fallen timber would restore the instream habitat.

Conclusion

In New South Wales, the *Biodiversity Conservation Act 2016* and the associated *Biodiversity Conservation Regulation 2017* provide a formal framework to avoid, minimise and offset impacts on biodiversity from development and clearing through the Biodiversity

Offsets Scheme. Although focused on mitigation, and typically on terrestrial-based proposed developments, there is some scope for implementing biodiversity offsets outside of the scheme. In most instances, however, there is, at best, limited scope within the legislation to seek to offset previously degraded habitat that is outside the scope of a proposed development. Community-driven restoration of environments (the topic of this chapter) does not fit within the thrust of current offset legislation and nor was it meant to do so – the legislation has a different focus. Likewise, the legislation pertaining to recreational fishing focuses on the resource – fish for recreational fishing – rather than on restoration of the degraded habitat, notwithstanding the importance of a healthy habitat in terms of fish populations.

Undoubtedly, there is great pressure on the habitat of freshwater fishes, particularly in expanding urban areas. With the plight of species, such as the Australian bass, that are suffering increasing recreational fishing pressure and continued erosion of the quality of their habitat, it is necessary to offset the past damage and ongoing degradation, regardless of this being outside the current offset-based legislation. There is a need for organisations (in this case Recfish West) to make their case and, hopefully, be listened to.

'Roof gardens': an emerging offset opportunity

Craig Langston, Kayalvizhi Sundarraj Chandrasekar and Shelley Burgin

Relentless expansion of urbanisation is a contemporary matter that requires a balance between development and retaining the natural environment. Too often this involves trade-offs between one and the other. But what if both could go some way to co-existing? This chapter contemplates the use of green (or living) roofs as a means of (partly) offsetting the loss of natural areas to urbanisation. We recognise that even a city of living roofs is not like-for-like compensation for the loss of a natural ecosystem. However, as any gardener with a liking for birds and bees knows, planting the appropriate flowers and shrubs will attract them – small compensation, admittedly. There are other values of a green roof, as we will consider.

Do green roofs have the potential to gain in popularity and, as with rooftop solar panels, make a contribution to environmental offsetting in the built environment? Green roofs might be the future. They were in the past! In the Viking era, communal dwellings were covered by thick grass. The grass played a crucial role as insulation in the cold and damp climate of Scandinavia.

Background

Australia is unique in that its national parliament has a green roof – it is covered in grass, on which citizens and visitors have open access to walk. Some compare it to a British lawn and do not see it as quite appropriate for Australia. Others express the view that it allows citizens to walk on top of their politicians. Be that as it may, Parliament House is a superb building of 4700 rooms, designed by Italian architect Romaldo Giurgola and opened in 1988, and the green roof provides additional open space around the building.

Based on the trends in recent years, the *State of Australian Cities* report (DIRD 2013) predicted that heat-related deaths are highly likely to increase, especially in Perth and Brisbane. Consequently, the liveability of Australian cities is dependent on how these cities are managed in a warming climate – and not just for humans. Strategies that focus on planting and, where it exists, retaining vegetation including shrubs and trees, extending to large green belts and urban forests, are essential to mitigate the urban heat island effect and related issues, such as energy demand for air conditioning in urban areas. Green roofs could play a role in modifying temperatures in Australian cities, and add to the ambience of these urban areas.

This is because roofs are generally one of the hottest surfaces of buildings during day-light hours. Vegetated (green) roofs have the potential to mitigate surface temperatures (Norton *et al.* 2015). But this is yet to be proven at a city-wide scale. However, modelling studies have shown a positive influence in terms of extenuating the urban island effect (Rosenzweig *et al.* 2006; Gill *et al.* 2007). As an aside, such an 'island' is an area, usually the inner city, that is hotter than the surrounding vegetated areas. For example, in a city of 1 million people or more, the inner city with extensive areas of hard surfaces may be 1–3°C hotter than its surroundings (EPA undated). A potential benefit could therefore result from widespread construction of green roofs in such urban areas.

In a position statement advocating urban green infrastructure, the Australian Institute of Landscape Architects (AILA 2016) envisioned a network of planned and unplanned green spaces and places, spanning both public and private realms of urban areas, that would provide an integrated system of benefits. As a complex, such green infrastructure includes trees in clusters as a small forest or individually planted throughout urban areas, including within central business districts where the heat island effect is often most acute. Such infrastructure could be incorporated into any grassed, or otherwise vegetated, open spaces. It could also include plantings, for example, on roofs and vertical facades – the topic of this chapter. Such vegetated areas could make a valuable contribution to an urban landscape in any geographical and climatic setting.

Green roofs have the potential to be of significant benefit to Australians since a par-ticular feature of the population is that we concentrate in urban areas. Indeed, most Aus-tralians – in the order of 90% – live in cities and towns. The suburbs in these urban centres sprawl due to the availability and cost of land, and a cultural predisposition for backyards. This is unlike most large European cities which typically have a high concentration of resi-dents occupying apartments in or near the central business districts, with effectively no open space other than local parks and gardens.

At present, however, there is limited demand for green infrastructure for roofs, walls and other vertical/semi-vertical surfaces in Australia. Consequently, the local construc-tion industry has not had a compelling reason to develop a business interest in greening buildings. However, opportunity knocks – all Australian cities have central business dis-tricts with an increasing concentration of high-density residential unit blocks replacing the single-family homes of the past. The typical Australian backyard, garden and rotary clothes-line are now rare in such areas (Burgin 2018b). Cheek-by-jowl roofs, separated by roads and footpaths, combine with towering walls to form what some unkindly call 'con-crete jungles'. There would seem to be a case for greening such areas with green roofs, walls and any other surfaces.

What is a green roof?

As an aside, it is necessary to define a few technical terms. First, a bit of jargon. Integrating natural features, such as green roofs and walls, as functional design elements of built struc-tures is referred to as biophilic urbanism. Broadly, there are two types of green roofs, tech-nically classified as intensive or extensive structures. Extensive green roofs typically have a soil thickness of less than 15 cm and are non-trafficable (not suitable to walk on), while intensive green roofs have more than 15 cm of soil thickness and are designed to be accessible.

The choice of green roof type is influenced by the load-bearing capacity of the underly-ing structure. Clearly, the capacity would influence the type of green roof that could be

Table 7.1. Green roofs in Australia

Urban infrastructure	Semi-intensive/ semi-extensive green roofs	Extensive green roofs
Public spaces		
Parliament House, Canberra	Museum of Old and New Art, Tasmania	Venny Community Centre, Melbourne
University of Melbourne South Lawn car park		Victoria Desalination Plant, Wonthaggi
National Gallery of Victoria international sculpture garden		Toilet block, Beare Park, Sydney
Crown Casino, Melbourne		Kingston School, Hobart
Freshwater Place, Melbourne		Children's Hospital, Brisbane
Minfie Park green roof, Melbourne		
Office buildings		
30 The Bond, Sydney	Council House 2, Melbourne	Pixel Building, Melbourne
Queenscliff Marine and Freshwater Discovery Centre		
Adaptive reuse		
		Waverton Coal Loader, North Sydney
Residential buildings		
M Central, Sydney		

fitted (or retrofitted) and, indeed, influence the structure of new builds where the owners seek to incorporate a green roof. In Table 7.1, we provide a list of what are commonly called green roofs (Loh 2009) – the list includes roofs that are classified as brown roofs. These have a coating of gravel and/or recycled material base with a thin soil cover. While it may be possible to grow green plants on a brown roof, opportunities are often limited with such a construction.

Within the classification of green roofs, there are wetland green roofs. One potential water source for water plants grown in such a situation is recycled grey water. Another green roof type may include potted plants in any variety of containers, and a wide variety of plants. A roof could be converted into food-production and even include hydroponic food production.

In addition to brown and wetland green roofs, a further sub-classification is known as a cool roof. This green roof sub-category is sometimes used as an all-embracing sub-classification that encompasses white (reflective), green (living) and blue (liquid) roofs. Despite the differences in make-up, all these sub-categories of green roofs typically have one common purpose – reducing the temperature within the urban heat island. However, many are additionally valued for their aesthetic and/or functional values.

Green roofs typically comprise a waterproofing barrier with a root barrier, drainage layer, filter membrane such as geotextile cloth, growing medium (soil layer) and vegetation. Insulation and its placement varies in accordance with climate and government regulations. Irrigation systems may be added if the focus is on growing plants. These layers all sit on a solid base, typically concrete for intensive roofs, but often lightweight structures such as plywood on timber or steel framing for extensive roofs.

It has been calculated (Oberndorfer *et al.* 2007) that green roofs could retain 25–100% of the rainfall depending on the slope of the roof, soil and root depth and, obviously, the quantity of rainfall. Shuster *et al.* (2008) found support for this finding. They observed that green roofs could effectively delay peak run-off, thereby reducing stress on urban storm-water infrastructure. Although the same could be said of bio-retention filters and rain gardens, green roofs offer additional benefits such as energy savings by reducing indoor temperature and increasing visual amenity.

Installing green roofs improves air quality and introduces an element of biodiversity that is very much city- and context-specific. In regard to the urban heat island effect, Norton *et al.* (2013) found that the performance of green roofs in reducing surface temperature is at a maximum when there is high and dense vegetation cover with large leaf spread at different heights. That is, intensive green roofs are found to be more effective than extensive roofs. However, most green roofs in Europe and North America – Toronto, Stuttgart and Berlin are the front-runners – are of the extensive type that can support only small plants or ground cover. In this case, sedum, which is a hardy succulent of the Grassulaceae family, capable of forming a cushion-like ground cover, is the most popular plant choice due to its drought-resistant characteristics.

Green roofs have many roles

It is worth noting that when the post-World War II government of Germany took to rebuilding its bombed cities, it mandated the incorporation of green roofs as a design element. Not every roof was to be covered in grass or shrubs, but the concept caught on and eventually a green roof movement spread widely to the rest of Europe and North America.

Green roofs are not limited to commercial and residential roofs. What we would now consider to be green roofs have been around in Australia for a long time under different guises – rooftop gardens, earth-sheltered buildings or simply landscaped plazas. Their primary form of use has been to cover underground garages to form parks. Brisbane's Southbank Parklands' intensive roof, including swimming lagoons over car parking structures, is a typical example. Other examples are the South Lawn at the University of Melbourne Parkville campus, and the outside, grassed sculpture area at the National Gallery of Victoria.

Greening Australian cities: not all plain sailing

When it comes to green roof adoption, Melbourne and Adelaide play lead roles in Australia, particularly underpinned by a focus on addressing climate change. Leaders of these cities are working towards achieving carbon-neutral cities. However, it is not appropriate to single out just those cities. Australia's capital city, Canberra, is rapidly moving towards carbon neutrality in electricity use. Solar and wind farms, with a total capacity to offset the greenhouse gases produced by the old coal-fired generators, are being built. The residents of Brisbane are world leaders on a per capita basis in installing rooftop photovoltaic electricity-producing panels. Also in Queensland, one Sunshine Coast local government generates sufficient power from a solar farm to cover all the local government electricity requirements. However, rooftop solar panels were ruled out in that instance because much of the local government's infrastructure (e.g. swimming pools – its largest consumer of power) had inadequate roof area to place sufficient solar panels, or indeed green roofs, to offset electricity consumption (Chapter 8).

Most of the research and demonstration projects undertaken in Australia are based on extensive, rather than intensive, green roofs. However, as Razzaghmanesh *et al.* (2014) pointed out, there have not been studies to determine the performance of intensive green roofs in Australia, especially in drier areas. Williams *et al.* (2010) did conclude that growing plants on roofs in the Southern Hemisphere is a different proposition from doing so in the temperate, colder and wetter climes of the Northern Hemisphere. Although there is little concern for plants needing to tolerate freezing winters across most of Australia, any rooftop planting would need to withstand temperature extremes which have been increasing with climate change. Plant selection and watering are crucial factors in maintaining roof gardens. It is obvious that location-specific and species-specific research is needed.

Innovations in green roof applications

Householders growing some of their food requirements in backyard gardens, and establishing fruit trees such as custard apples, mulberries, bananas, mangos and paw-paws, has a long history in Australia stretching back to the first European arrival. However, the importance of the home garden to supplement the household budget has dwindled. The greatest change has occurred since World War II. Home food production is an unusual practice today, especially in families where household adults are in employment and/or studying, and with the financial resources to buy from supermarkets and other outlets that offer an extensive range of fruit and vegetables. Nevertheless, with the rise in interest in healthy lifestyles (e.g. organic foods) home-grown produce is increasingly preferred by some (Burgin 2018a). Some consider that home-grown food tastes better, and is healthier without the addition of chemicals and artificial fertilise. Rooftop gardens should appeal to some of these people, although the associated costs are likely to be an impediment to implementation.

Research is being undertaken in paddy fields and natural wetlands on plant microbial fuel cells as an alternate technology to enable production of electricity from living plants. Such research could offer a new sustainable and non-polluting source of electricity generation (Wetser *et al.* 2015). Helder *et al.* (2013) discuss the potential for using similar technology on rooftops. A pilot project on a green roof has been successfully demonstrated in the Netherlands.

Exemplar green roofs in Australia

One of the best-known examples of a green roof in Australia is Parliament House in Canberra. However, it has received its fair share of criticism. For example, Graeme Hopkins, a South Australian expert on green roofs, criticised it for not being much more than an outdated replication of an English lawn. He suggested that the roof would perform well in reducing the carbon footprint, if the lawn was removed and native succulents and wild flowers from the Snowy Mountains (an area broadly within the same region as Canberra) were planted instead of the water-intensive turf (cited in Beeby 2012).

On the other hand, the Queenscliff Marine and Freshwater Discovery Centre in Victoria is seen as a successful paragon – a poster child – of ecology and design. The stunning structure enables the occupants to be comfortable without air conditioning, due to the ambient temperature facilitated by its green roof (Dillon 2008). A substantial number of buildings, predominantly public buildings, also sport green roofs (see Table 7.1 for some of the better known).

Inspiration for Australia's largest green roof

A US example was the catalyst for the Wonthaggi green roof. The largest living roof in the world sits over the Ford Motor Factory in Dearborn, Michigan. This green roof demonstrates how creative thinking can transform an environmental issue into sustainable practice. The roof is ~10.4 acres (slightly less than 25 000 m²). It is covered with sedum and other succulent plants.

Compared with a conventional roof, rainwater run-off from the Ford Motor Factory's green roof is substantially reduced. What run-off there is would be slowed since the roof can hold ~1.3 cm depth of rainwater. Additionally, the plants require carbon dioxide to grow, and thus sequester it from the surrounding air.

The Wonthaggi desalination plant, Australia's largest green roof (26 000 m²), is set in 255 hectares of coastal lands – and vegetated coastlines have a special appeal if they appear natural. It was designed with the objective of reducing the visual impact of the plant. By creating a green roof, the desalination plant roof blends into the surrounding landscape, particularly since the plant species grown on it are present in the local landscape (Gledhill 2011).

Two unique green roofs adorn other buildings in southern Australia: the Museum of Old and New Art – commonly referred to as MONA – in Hobart, and the Pixel Building in Melbourne.

Deceptively small from the outside as it is largely built into rather than above the ground, MONA has a huge and absolutely eclectic collection and is a major drawcard for visitors to Hobart. It is covered by 1400 m² of green roof space, designed specifically for use as a public open space (Nicholls undated).

The green roof of the Pixel Building has a very different focus from the landscaped garden space of MONA. It was the first carbon-neutral building in Australia, and received the highest green building rating in the world in 2010. The building has an extensive green roof, planted with native grasses which play a key role in rainwater collection and filtering.

However, buildings do not necessarily have to be purpose-built as green-roofed buildings, so we now offer descriptions of some old buildings that have been redeveloped with an eye to the future.

The first is the Coal Loader Centre for Sustainability in the Sydney suburb of Waverton, which involved redevelopment of an old coal loader platform as part of the Waverton Peninsula Strategic Master Plan (North Sydney Council 2020). The 1 hectare green roof over the loading platform includes urban harvest plots of vegetables and fruit trees.

The second repurposed green roof building we would like to introduce is a late 19th-century wool store, although the supports for the green roof are not an original part of the wool store. The building was retrofitted in 2005, and the constructed 2600 m² roof garden is used as a recreational space on a regular basis by at least 200 of the building's 400 residents (Angel 2014). It is not unique for wool stores to be repurposed in Australia – there are very fine examples of repurposing of wool stores into upmarket apartments in Teneriffe, Brisbane.

In the past, retrofitting has typically stopped short of incorporating roof gardens. However, there is an outstanding roof garden at the Children's Hospital in Brisbane.

The Children's Hospital (Fig. 7.1) is situated in a relatively high-density area of South Brisbane –not always the most salubrious address in Brisbane! The total area of roof garden

Fig. 7.1. Part of the roof garden of the Children's Hospital, South Brisbane. Photo: Craig Langston.

is ~3200 m². On this roof, there are some 46 000 individual plants, eight shelters, 12 monoliths and 33 columns covered in epiphytes (plants that grow on a diversity of surfaces and derive moisture from the air, rain and debris that accumulates around them). Additionally, there is a sloping green roof with 1400 planting cassettes containing 23 000 plants.

Conclusion

With the limited popularity of roof gardens in Australia at present, their role in terms of environmental offsets is minimal. However, with greater popularity and the increased coverage of buildings, there will be greater benefits including urban heat island mitigation, reduction in energy demand, stormwater management, visual amenity, recreation, food production and carbon dioxide reduction. With improvements to design and construction, as occurred with the increasing uptake of rooftop solar energy panels in Australia, we predict that in the future rooftop gardens will become increasingly popular and will take their place in the mix of environmental offsets.

8

Offsetting coal-based electricity

Tor Hundloe and Keeley Hartzer

Queensland deserves its title as the Sunshine State. North of the state's capital city Brisbane, roughly an hour in travelling time, is what is formally called the Sunshine Coast. It has a population of about one-third of a million. Local government matters are dealt with by an elected council. Local issues include maintenance of minor roads and footpaths, operating local libraries and swimming pools, recreational parks, caravan/camping grounds, vehicle parking, and garbage collection and disposal, plus a long list of other minor matters which Australia recognises are best dealt with at the local level.

On the basis of population, the Sunshine Coast (as a city) is ranked ninth in Australia. Its considerable emission of greenhouse gases was recognised by the local politicians, who decided to offset the greenhouse gases resulting from the local government's use of electricity. This was a significant task. As there was not enough land available locally to plant a very large forest to soak up carbon dioxide, that form of offsetting was not available. Their solution was to stop emitting greenhouse gases while operating council facilities – yet, of course, electricity was necessary. This meant constructing a solar or wind farm. Other sources of renewable, carbon dioxide-free electricity such as hydro and wave power were ruled out on practical grounds.

The Sunshine Coast made history in mid-2017, becoming the first city-sized local authority in Australia to run all local government operations, excluding council vehicles, on non-fossil fuel-based electricity. A solar farm had been considered the best option, if sufficient vacant land was available – this became the major issue. For those who may consider that building solar farms is no problem in a country the size of Australia, there are a fact or two to bear in mind.

Australia is a vast country, approximately two-thirds of which is desert or semi-arid, therefore it has plenty of room for extensive solar farms without disturbing any productive industries – at least in remote areas. But the Sunshine Coast is not an outback town surrounded by very low value arid land. Coastal land tends to have high economic value, which can very easily rule out a solar farm of any size.

Residents of the Sunshine Coast are overall environmentally conscious, not only in supporting the concept of the solar farm but in protecting the natural beauty of their city with its extensive beaches abutting the Pacific Ocean and its rainforest hinterland. As a consequence of this pro-environmental attitude, there are limits on building heights, construction on the intact frontal dunes is forbidden and there are very few canal housing estates that eat into mangrove ecosystems.

The Sunshine Coast comprises golden beaches, frontal dunes in the areas where wise governments intervened before the developers had their way, coastal floodplains along rivers that wind from the hinterland to the ocean, some cleared farming land with a variety of uses (mainly sugarcane growing), and finally the mountainous rainforests to the west.

As circumstances change, land becomes available

Sugarcane farms (plantations) need to be relatively close to a sugar mill to reduce the cost of transporting the cut cane, which is normally carried by miniature trains running on specially laid lines through a farming district. Sugarcane-growing districts developed around this concept. If there were enough sugarcane farmers in an area, a mill was built. This was the case at the close of the 19th century when the Sunshine Coast floodplains attracted farmers. The Moreton Sugar Mill, in the centre of the small Sunshine Coast town of Nambour, opened in 1897. The local sugar industry remained buoyant for a century.

The last cane train to deliver cut cane to the Moreton Sugar mill rolled into town on 3 December 2003. From that date on, sugarcane farmers in the Sunshine Coast district were forced to transport their cane 160 km to the mill at Maryborough by road transport. The added cost affected the financial viability of many farmers, and the 2004 cane-growing season resulted in numerous bare-land farms. What was to become of the land?

The future use of this former sugarcane land was not just an economic and environmental issue, it was political in the sense that the local community was interested. Local politicians had to reconcile conflicting land use proposals. In 2009, the local council produced a *Canelands Discussion Paper* for community consultation. A solar farm was not on the agenda at that stage and hence was not mentioned. However, the idea of building one was gelling. The council also commissioned a study called *Regional Energy Opportunities*. This was key to opening up consideration of a solar farm – local government officials were seeking forward-looking strategies.

In mid-2010, the Sunshine Coast Council adopted a Climate Change and Peak Oil Strategy, a far more radical policy initiative than those produced at the higher levels of government in Australia (state or Commonwealth). The notion of peak oil was not then, and still is not, on most government agendas in Australia. History should therefore record this as a revolutionary project at local government level in Australia. At the end of 2010, the council adopted its Energy Transition Plan and history was about to be made. The plan allowed renewable energy production as a complement to farming on rural land. It opened up the question of whether good quality agricultural land should be made available for conversion to a solar farm and if so, on what conditions?

Here we temporarily leave the dilemma faced on the Sunshine Coast, and turn to the national level. The question of determining on a national scale what land is suitable for either solar or wind farms was addressed in 2012 by the Australian Energy Market Operator. The results of the study were, in general, practical and sensible. For example, grazing land, comprising both natural and improved pastures, was deemed suitable for installing wind farms due to their very tiny footprint on the large properties; natural grazing land (of which Australia has an enormous amount) was deemed suitable for solar farms; forestry areas were ruled out for obvious reasons; and cropping and horticulture land was not entirely excluded. All land used intensively was considered not suitable, including land used for intensive horticulture. Under this assessment, existing sugarcane farms (intensive farming) were not to give way to solar farms. Retired sugarcane land, being no longer productive in its original use, was not specifically considered – the Sunshine Coast Council could make its own determination.

A solar farm is built: a preventive offset

A vacant sugarcane farm became the site of the first solar farm in Australia that had been developed by a local authority, and when it started operating in mid-2017 it was the fifth largest solar farm in Australia. Named Valdora, the solar farm is sited on a 49 hectare block of land situated between the Maroochy River and the Yandina-Coolum Road.

Today, approximately 20 hectares are covered with 48 000 photovoltaic solar panels, which are raised above the flood level as this is floodplain country. The solar farm has the capacity to provide 15 MW of electricity – enough to power all the Sunshine Coast Council's administration buildings, aquatic centres (swimming pools), community venues, sporting venues, holiday parks, libraries and art galleries. The amount of electricity is equivalent to that required for 5000 households. The electricity is fed into the grid network sub-station on the border of the solar farm, not directed to council facilities. This close proximity to the grid is a significant economic benefit.

It could be asked – why did not the council simply install photovoltaic solar panels on its buildings and service centres? At the time that the Valdora project was being planned there were already 30 000 rooftop solar arrays on residences and small business premises on the Sunshine Coast. The number has increased substantially since then. The council's problem was that its roof space was limited. For example, the numerous public swimming pools and aquatic centres, which are the main users of electricity, have virtually no roof space on which to install solar panels.

There was some public opposition to the development of the solar farm, including legal challenges which the local council won. Aesthetics and glare were the issues of concern. Residents in distant hillslopes do have a partial view of the site, but distance and vegetation screening overcame the problem.

The capital cost is reported to have been AU$50.4 million. The net benefit savings in electricity bills for the council plus money earned from the excess electricity sold into the grid has been estimated at $22.1 million over 30 years, based on present electricity costs. The Valdora solar farm is expected to save over 20 000 tonnes of carbon dioxide annually, equating to approximately 40 000 tonnes of coal not burned. Consequently, there is a significant reduction in greenhouse gases compared to the situation before the solar farm was built. The benefits are therefore two-fold: reduced electricity bills and reduced greenhouse gas emissions. Whether other local authorities will follow suit, only time will tell; for example, not many will have the opportunity to use vacant flat land close to a grid sub-station.

If we make assumptions about the damage avoided by reducing greenhouse gas emissions, we could put an economic value, over and above money earned from sales of excess electricity, on the solar farm. If we assume the carbon tax that was in place in Australia for a short period (set at AU$23 per tonne of carbon) is a reasonable estimate of the damage per tonne of carbon emissions, the result is that just under AU$500 000 in damage is avoided each year into perpetuity (if the solar farm is maintained indefinitely). We could use a much higher carbon price. For example, in 2006 Nicholas Stern suggested a range of US$25–85 per tonne. Based on the high estimate, the value would be US$1.7 million annually.

Fig. 8.1.   University of Queensland solar farm – Gatton. Source: © University of Queensland.

Conclusion

All in all, the Valdora solar farm is impressive. It is comparatively small compared to the ones that are being constructed as we write but it is multi-purposed – the Sunshine Coast Council also plans to run sheep on the site. The Australian model for this is the University of Queensland's very small solar farm at Gatton (see Fig. 8.1).

CASE STUDIES: OFFSETTING IN DEVELOPING COUNTRIES

In this section of the book we briefly consider the use of offsets at the international scale, then focus on the trials and tribulations of developing countries seeking to use the offset concept to achieve conservation goals.

Offsets in developing countries

Tor Hundloe

With regard to the role of offsets at a global scale, the focus is on how a poor country can be compensated for a pro-environment action. For example, we want the planet's remaining rainforests to be preserved – how can we achieve that? In poor countries, forests are felled legally so the land can be used for growing crops or raising cattle in order to earn much-needed foreign exchange. There is also considerable loss of forests from illegal logging.

With regard to legal forest clearing, there is a significant opportunity cost involved in the preservation of forests – the much-needed but forgone agricultural income. Soy production now takes up large areas of previously forested land in South America, while palm oil production does likewise in some South-East Asian countries. Offsetting the income not earned if nature conservation is to take precedence over large-scale farming is one of the most urgent issues in environmental management in the poor countries. It is of global concern, and this puts the onus on the rich countries to help out. The seemingly ever-increasing consumer demand for soy-based products and for palm oil in its variety of uses is driven by their use in the richer countries and by population growth in the developing countries.

It is not that efforts have not been made to reduce forest clearing due to illegal activities, but as forests continue to be felled at an alarming rate, the effort has been far from enough. There would seem to be little interest in addressing the growing demand for soy and palm oil. It should be noted that the rich countries can offer considerable aid but have little in the way of positive results if the recipient government is not committed to preserving its forests.

In terms of illegal logging, the issue is a lack of resources for surveillance, monitoring and enforcement of the law. In vast rainforests, such as the Amazon, on-ground surveillance has proven inadequate – very inadequate. The situation is not that different in parts of South-East Asia, sub-Saharan Africa and certain South Pacific island states. Recently, surveillance in Brazil has benefited by the donation of an aircraft by an environmental non-governmental charity.

Ever since the first UN conference on the environment, held in Stockholm in 1972, the rich countries have been aware that they have to help the poor countries if global environmental problems are to be solved. As concern about climate change took hold in the late 1980s, the rich countries have, either unilaterally through their foreign aid programs and/or by offering assistance via issue-specific agencies of the UN (in particular the UN Framework Convention on Climate Change), increased their efforts to assist.

For some decades now the richer countries of the world have provided various incentives to the poorer countries to protect important natural ecosystems, particularly rainforests. One way in which this has been attempted is by agreeing to forgive some of a poor country's foreign debt if that country is willing to protect certain natural resources. These debt-for-nature swaps became popular when international rock stars promoted the concept. However, this approach has diminished in importance.

Commencing with the first serious attempt to control climate change, the Kyoto Protocol in 1997, the UN initiated its Clean Development Mechanism. This works through a formal agreement by which a rich country funds a project (for example, an afforestation or reforestation project) in a poor country – thereby creating a carbon sink – and accruing certified emission certificates which help to meet the rich country's agreed greenhouse gas emission reductions.

The most recent program goes by the title REDD+. This scheme offers incentives to the poor countries, formally known as developing countries, to protect forests from deforestation and degradation. The goal is to reduce the increasing greenhouse gas emissions.

We might expect that, given this effort and others, the major, globally important forests would be protected. This is far from the case. It is unusual for *The Economist* to lead – and herald its leading story on its cover – with an environmental issue, but that is what it did in the edition 3–9 August 2019. 'Deathwatch for the Amazon' the cover asserted, as only covers can do. The article inside deserves quoting. It commences with a global overview and discusses the denial of science and planetary responsibility by Brazilian president Jair Bolsonaro, who incidentally has threatened to leave the Paris Climate Agreement:

> *In the tropics, which contain half of the world's forest biomass, tree-cover loss has accelerated by two-thirds since 2015; if it were a country, the shrinkage would make tropical rainforest the world's third-biggest carbon-dioxide emitter, after China and America. Nowhere are the stakes higher than in the Amazon basin – and not just because it contains 40% of Earth's rainforest and harbours 10–15% of the world's terrestrial species. South America's natural wonder may be perilously close to the tipping-point … There are hints the pessimists may be correct … Brazil's president dismisses such findings, as he does science more broadly … 'The Amazon is ours', the president thundered recently …*

The implication: 'we will do what we – I – want with it.'

If we note the various attempts to help Brazil – to compensate the country if it protects its vast rainforests – we must conclude that whatever has been tried has not worked. However, there have been promising signs in the past. The international Amazon Fund, created in 2008 to help pay for protection, slowed the rate of deforestation. It did not last, notwithstanding the REDD+ scheme or anything else. The most recent donations to Brazil for forest protection come from Norway (a country with a tiny population compared to Brazil) and Germany, in the amount of US$950 million!

Debt-for-nature swaps and the various climate change-related schemes are not strictly offsetting arrangements as usually defined, but they are a means of compensating a poor country if it takes a pro-environmental position and deserves help and compensation. The evidence to date is that we have made virtually no progress on what is called the 'north–south divide' on climate change, protecting biodiversity and fair trade. The latter would be a significant help to the poor countries, in terms of both environmental protection and economic development.

Who is responsible for environmental damage and who should pay the price?

There is a significant matter in terms of global responsibility for polluting activities, which results in confusion and buck-passing. The issue is simply this: if country A is a producer and exporter of coal and country B is the importer of that coal and burns it in furnaces to produce electricity, who is to blame for the resulting greenhouse gases? Is country A or country B responsible for offsetting the carbon dioxide and other greenhouse gases?

There is an agreed international principle which deals with this matter. However, not all interested parties agree, or at least feel comfortable, with the principle as it stands. So, what is the principle?

The responsibility to offset pollution from the consumption or other use of an imported product rests with the people in the importing country. We can make this clear with an example. Australia is a major exporter of coal, gas and uranium. The burning of coal and gas results in the emission of greenhouse gases. The offsetting of these emissions as the electricity is generated (if that is the purpose of importing these fossil fuels) is the sole responsibility of the ultimate consumers, whether they be households or industries in Korea, India or anywhere else.

The principle underpinning this rule is what economists call 'consumer sovereignty' – in other words, the consumer is deemed king or queen (sovereign). The producer is doing their duty to respond to the 'sovereign's' command to produce a good or service. This proposition is the philosophical justification for our present global economic system. If the consumer is sovereign, the consumer pays – not only the price to purchase the product but for any pollution caused by using the product.

There are two exceptions to this principle, which illustrate that there is no consistency or jurisprudential thought in the matter. The exceptions are not universally applied, which is a further concern. The exceptions to the consumer as sovereign rule are prostitution and drug-taking. In both cases, the supplier of the service or substance is prosecuted and the consumer usually either goes free or is treated far more leniently.

If coal is mined in Australia and exported, under this principle, the consumers in the importing countries are responsible for offsetting the carbon dioxide. We can understand the consumer sovereignty principle as applied to pollution. – the polluter is the consumer and it is their behaviour that we seek to change. Hence, the principle that 'the polluter pays' makes sense. However, there is at least one practical reason to put the onus on the producer rather than the consumer, and it is simply that there are only a small number of producers in several important industries (e.g. electricity generation, steel-making) but millions of customers of the final product. Both administrative costs and ease of surveillance are diminished if only a few businesses are targeted. This is why carbon taxes and 'cap and trade' schemes focus on the major greenhouse gas-emitting industries, rather than on the millions of final consumers.

The Australian government's position is shown in the context of coal exports. Given the polluter-pays convention, there is no requirement to offset in Australia the greenhouse gas emissions which will result from burning the Australian coal overseas. Any offsetting is to be undertaken in the importing countries. The Commonwealth Department of the Environment and Energy makes this clear as follows:

With regard to the impacts of the emissions caused by the use of the coal … recipient nations will need to meet their obligations under the United Nations Convention on Climate Change (DOE 2015).

It is possible to change this principle, notwithstanding its strong philosophical position. Governments, acting both unilaterally and in common, can change economic principles that are far less binding than the laws of nature that King Canute attempted to control.

Case studies: Cameroon and Tanzania

The next two chapters focus on examples of offsets that are the sole responsibility of the country concerned. This does not mean that foreign governments, multi-lateral organisations, environmental non-government organisations and others do not play a role in providing resources or expert advice or in any other way assist the poor country to achieve a desired environmental outcome. Many poor countries are forced to rely on such support to achieve a desired conservation result.

Our focus is on the challenges a poor country faces in offsetting a pro-environment initiative, such as declaring a major national park or preserving historical buildings. Doing this can be a significant political, economic and cultural challenge even for the richest countries in the world – the opportunity costs can be high, such as with the forced cessation of forestry, mining or other profitable land uses so that national parks or World Heritage areas can be declared.

Australian environmental history is replete with examples of the political – and in many cases moral – imperative to compensate the losers from a pro-environmental action. We discussed this type of compensation as a form of an offset in Chapter 1. Pro-environment initiatives are difficult in the rich countries, but they are extremely difficult in the poorest of countries. Feasible offsets are not easy to identify and, if identified, implementation is an economic challenge.

It is worthwhile to distinguish the different situations where offsetting is required. The usual case is where a development project such as an airport is to be built, and notwithstanding our best efforts there will be residual negative impacts. For example, some hectares of mangrove forest will be lost. The same area of forest would need to be replanted as close as possible to the lost forest, with the aim of there being no overall negative impact on mangrove-dependent fisheries. Even then the issue of the delay between removal of the original trees and maturity of the replacement ones – and hence production of the same type and amount of ecosystem services for marine life – suggests a practical problem with which experts grapple.

If the project is to mine oil, we cannot replace the oil as we replaced the mangroves. Mining oil is an irreversible decision. How do we offset this? There are various alternatives. If we assume the oil is refined to become diesel or petrol – and to make the example very clear, let us assume we know that we are near the end of the global supply of oil – we could plant various bio-fuel crops to replace all the oil mined. Or, if this offset is not to your liking, we could speed up the development of hydrogen vehicles as replacements for the internal combustion vehicles of the present era.

The other major category of offsetting is relevant when we are dealing not with a development project but with a pro-environment project. We are going to stop an existing money-earning activity to preserve something we have come to value more highly. Let us say we are to preserve a historic monument or building – keep it for all time so that future generations can benefit from its existence. This, like mining oil, is an irreversible decision. In this case, offsetting requires compensating the owner of the building if they want to exercise a right to demolish it and construct a building which would generate greater income for the owner.

The case studies we present in the following chapters are of the pro-environment kind. There are opportunity costs involved because near-natural environments and historic buildings are to be preserved for public enjoyment and appreciation, and for the potential to draw foreign tourists and hence produce income for the nation. These places and buildings could otherwise be destroyed by their owners or users in pursuit of personal economic gain or continuance of the lifestyles of those using them today.

Offsetting, in various forms of compensation, is required if environmental outcomes are to be achieved. Because we are going to focus on poor countries – we use the terms 'poor' and 'rich' rather than official UN descriptions because we prefer our language to be truthful – the situation is not one where it can be assumed that these countries' governments, using taxes, can afford to pay the compensation. The challenge is to find other ways of compensating those who are required to give up existing benefits for the common good.

Selecting the case studies

Our initial task was to select a region, or regions, of the world best suited to explore offsetting pro-environmental initiatives. As we are to focus on poor countries, preferable very poor ones, Africa, in particular sub-Sahara Africa, picked itself. Most countries at the lowest end of the UN Human Development Index (HDI) are in this region. Depending on whom is doing the counting, there are either 46 or 51 countries in sub-Sahara Africa. The US Library of Congress lists 51, while the UN Development Programme lists 46. The latter is our preference. Of the 46 countries, only five are not very poor (Botswana, Equatorial Guinea, Gabon, the Seychelles and South Africa).

An indication of the material well-being of a country is the length of life of its people. The average in 2016 for sub-Sahara Africa was 59 years. The two countries with the lowest length of life are Swaziland at 49 years and Sierra Leone at 51 years, while the two with the longest lifespans are Cape Verde (74 years) and the Seychelles (73 years). These numbers illustrate that nearly 25 years of life are lost on average, if we take Australia as the yardstick where the lifespan is in the order of 83 years.

In a nutshell, sub-Sahara Africa is, in general, very poor. This is illustrated by income data, notwithstanding its shortcomings as a measure of well-being in societies where subsistence agriculture and the informal economy play very significant roles. Per capita Gross National Income (GNI) measured as purchasing power parity for the sub-Sahara Africa countries, excluding the high-income ones, averaged out at approximately US$3600 in 2016 (World Bank). Four of these countries had incomes between US$700 and US$800 (Burundi, Central African Republic, the Democratic Republic of Congo, and Liberia). If we were to include Somalia, for which we have no reliable data, its average per capita income is likely to be half that of these extremely poor countries.

The other important consideration in selecting our case studies was to have one that dealt with nature conservation and the other with built heritage conservation. On this basis, we selected Cameroon on the west coast of Africa for its nature conservation and Tanzania on the east coast for its built environment conservation. The latter choice could seem on the face of it counter-intuitive, a matter we will come to.

The selected countries are similar in certain respects. The 2016 HDI rank for Cameroon was 153 and for Tanzania it was 151. Their respective average incomes, using the method described above, was US$2900 for Cameroon and US$2500 for Tanzania. There was, however, a marked difference in lifespans – Tanzania at 66 years and Cameroon at 56.

The matter of a counter-intuitive choice of Tanzania: Tanzania is recognised both as the likely birthplace of our human ancestors, and for its wildlife. The Tanzanian mainland (previously known as Tanganyika) is known world-wide for the pioneering paleontological research of Louis and Mary Leaky, and their discovery of evidence of our human ancestors in Olduvai Gorge. In terms of wildlife there is Serengeti National Park (a World Heritage property), as well as other UNESCO-listed fascinating places: Mount Kilimanjaro; the Arusha region with Ngorongoro Crater and the Maasai people coexisting with endangered wildlife; Selous Game Reserve; and the rock art of the Kondoa District in the Great Rift Valley. There are two historic/cultural World Heritage sites in Tanzania. One is Stone Town of Zanzibar, the other is the Ruins of Kisiwani and Ruins of Songo Mnara. Our case study will focus on Stone Town and a non-World Heritage site, the Dar es Salaam Historic Center.

Cameroon has its own natural treasures listed as World Heritage properties – Dja Faunal Reserve and Sangha Trinational. The latter, as its name suggests, is shared with two other countries, the Central African Republic and Congo. Our Cameroon case study deals with an area of natural forest that was declared a Strict Nature Reserve – the highest category of UN protected area, severely restricting access to the area. In addition to its biodiversity importance, it has extremely high indigenous cultural value.

Our hope is that these case studies singly and in combination are representative of the difficulty of achieving pro-environmental outcomes where something – it might be hunting in a rainforest, or it might be to 'interfere' with development opportunities in historical urban settings – has to give way and be offset. The offsetting will not be like-for-like. We cannot allow the destruction of ancient forests or heritage buildings and expect that we can replace them, and we will come to see this puts a different perspective on offsetting.

Compensating for the loss of tribal lifestyle

Eric Fru Zama, Shelley Burgin and Tor Hundloe

A major challenge for developing countries is to establish and maintain national parks, World Heritage properties, biosphere reserves or any other formally protected area. There is, more often than not, a not insignificant cost involved. Declaring a protected area, particularly if managed for strict nature preservation (at the very high end of preserving biodiversity) is likely to result in the curtailment of some, many or all existing human uses of the area. How is the method and cost of compensation for those who are to be excluded from the area determined? How is it offset? The cost does not have to be monetary; in fact, the effect on lifestyles and cultures will, in many cases, swamp any monetary loss.

Furthermore, there is cost to governments to manage and ensure the integrity of the newly protected area. In this chapter, we illustrate these issues by reference to the declaration of a large protected area in Cameroon. Our aim is to illustrate the need to offset pro-environmental actions by compensating those who are forced to forgo existing material benefits and cultural attributes, for the sake of the common good.

Our case study protected area is the Bakossi Forest Reserve in the south-west of Cameroon. It was gazetted at the highest end of the IUCN spectrum (1a Category: Strict Nature Reserve; see Table 10.1). Areas under this category are protected explicitly for the preservation of the uniqueness of their ecosystems, the animal and plant species which inhabit the area, and geodiversity features that have formed without human influence. In this situation, they can be degraded or even destroyed by nothing more than 'light' human impacts (Bishop *et al.* 2004).

For this category of reserve, emphasis is placed on the preservation of ecological relationships by maintaining the landscape as close as possible to being undisturbed by humans. Human access, and thus traditional use of the land and its wildlife, is excluded for all except research purposes and environmental monitoring. Consequently, the Cameroon government excludes – or certainly seeks to exclude – the indigenous peoples from their traditional lands in Bakossi Forest Reserve. This, as will be explained, is a case where offsetting in an appropriate form is needed in terms of equitable treatment of the traditional land users, and for the long-term success of the government's brave, forward-looking decision. At this stage, it is necessary to set the geographical scene in terms of the physical and human geography of both Cameroon and Bakossi Forest Reserve.

Table 10.1.   The IUCN six categories of protected areas

Category	Type	Characteristics
Ia	Strict Nature Reserve	Biodiversity protection, and protecting geomorphological features. Emphasis on conservation values with human access and use strictly controlled and limited.
Ib	Wilderness Area	Protection of typically largely unmodified areas that retain their natural character without permanent or significant human habitation. Protected and managed to preserve their natural condition.
II	National Park	Protection of large-scale ecological processes, species and ecosystem characteristics of area. Also provides foundation for environmentally and culturally compatible spiritual, scientific, educational, recreational and visitor opportunities.
III	Natural Monument or Feature	Protection of specific natural attributes, including landforms, sea mounts, submarine caverns, geological features (caves) or living features (an ancient grove).
IV	Habitat/Species Management Area	Protection of priority species or habitat of which many may need regular active intervention to address target species and habitat requirements.
V	Protected Landscape	Areas where interaction of people and nature has produced an area of distinct character with significant ecological, biological, cultural and scenic value, and where safeguarding the integrity of this interaction is vital to protecting the area and associated values.
VI	Protected Areas with Sustainable use of Natural Resources	Conserving ecosystems and habitats, together with associated cultural values and traditional natural resource management systems. They are generally large, with most of the area in a natural condition, where a proportion is under sustainable natural resource management and where low-level non-industrial use of natural resources compatible with nature conservation is allowed.

Source: Burgin and Zama (2014). Owned by authors, published by EDP Sciences, 2014.

The nation of Cameroon

Cameroon is a west coast, central African country sharing its boarders with Chad to the north, the Central African Republic to the east, Equatorial Guinea, Gabon and Congo to the south, and Nigeria to the north-west. It has a total surface area of 475 440 km² (117 484 000 acres), and a population of 23.44 million in 2016 (World Bank 2017). Politically and culturally, the country enjoys a dual system, a legacy of its former colonial masters, France and Britain. The only other country that is in this situation is Vanuatu in the South Pacific.

Following independence in 1960, Cameroon was one of the most prosperous countries in Africa. However, a reduction in prices of its principal exports (petroleum, cocoa, coffee, cotton) in the mid-1980s, followed by the devaluation of the franc in the 1990s, led to a fall in real per capita Gross Domestic Product by more than 60% between 1986 and 1994. Fiscal deficits increased and foreign debt grew.

As noted in the previous chapter, Cameroon is a relatively poor sub-Saharan African country, ranked 153 in the 2016 Human Development Index, with life expectancy at birth of 56. However, it has oil reserves, gold, diamonds and sapphires. Its good-quality agricultural land permits it to grow a wide variety of crops for both domestic consumption and export (UNDP 2016).

Bakossi Forest Reserve

In the south-west of Cameroon, located at 4°44'34"N and 9°35'20"E and covering an estimated 5517 km² (1 363 280 acres, or approximately 2.5% of the nation) is Bakossi Forest Reserve, which was officially declared a Strict Nature Reserve in 2000. Figure 10.1 illustrates a special feature of the area. Funding for establishment of the Reserve was provided by the Swedish World Wildlife Fund. From 1956, part of the area had been a declared forest area. This was before Cameroon became a unified nation in 1961. The Bakossi National Park, within the Forest Reserve, was declared in 2000 and came into being in 2008. This was a very important initiative by the Cameroon government, due to concern about a significant loss of natural forests. The conventional wisdom is that the nation's forests decreased by 40% between 1990 and 2015.

The Reserve is covered with dense tropical rainforest and encompasses the largest area of cloud forest in West-Central Africa (Lambi and Ndenecho 2009). It is extremely rich in

Fig. 10.1. Waterfall, Bakossi Forest Reserve. Source: Nick Annejohn and Family, Wikimedia Commons.

biodiversity (Bishop *et al.* 2004). Mount Kupe is particularly biodiverse, with many primates, birds and plant species endemic to the area. Of particular importance are the primates. The area has the healthiest remaining population of the endangered *Mandrillus leucophaeus* drill (a primate that is closely related to the *Mandrillus sphinx* mandrill).

After wide-scale timber logging (and easy access to hunting grounds) was banned, many years had to pass before the population of drill started to recover. Today, under its new environmental status, only small-scale timber-getting to construct homes and make furniture is allowed. These activities are considered to have minimal environmental impact. However, rapid population growth in the area will eventually result in increased timber-getting to build houses (Ngea 2011). This illustrates the difficulty of pursuing nature conservation while allowing traditional land uses. It is far from being purely a Cameroonian problem, as many other countries face the same dilemma.

Animals

In addition to the endangered *Mandrillus leucophaeus* already mentioned, many other primates occur in the forest, including the Preuss monkey (*Cercopithecus preussi*), red-eared guenon (*Cercopithecus erythrotis*), greater spot-nosed monkey (*Cercopithecus nictitans*), and several species of bush baby including (*Otolemur garnettii*), collared mangabey (*Cercocebus torquatus*), chimpanzee (*Pan troglodytes*) and Preuss's red colobus monkey (*Procolobus preussi*). Of large animals, there is the African forest elephant (Ngea 2011).

In addition to the many primate species, there are 329 known bird species recorded from within the Mount Kupe area. These include the Mount Kupe bushshrike (*Malaconotus kupeensis*), the endangered white-throated mountain babbler (*Kupeornis gilberti*), the vulnerable green-breasted bushshrike (*Malaconotus gladiator*) and the grey-necked picathartes (*Picathartes oreas*) (Ngea 2011). There are likely to be more bird species which are yet to be identified. There are as many as 2500 plant species of which 82 are endemic to the area, and 232 are listed in the Red Data List as threatened with extinction (IUCN 2012).

Clearly, Bakossi Forest Reserve is an area of extreme conservation value!

People

The Bakossi tribe, the traditional people living in and using the lands of the Bakossi Forest Reserve, number ~200 000 people. Other minor tribes live in the Reserve and there are immigrant people, the Bamiliki. The forest dwellers are mainly subsistence farmers and hunters (Fernando and Metuge 2016). Farming is very productive due to the rich volcanic, fertile soils. Cocoa and coffee are grown as cash crops, while plantain, bananas, coco yams, corn, cassava, beans, groundnuts and pepper are grown for home consumption with any surplus sold. Beef cattle are grazed but bush meat remains the traditional source of animal protein for the people. Bush meat is sold in the villages and cities of Cameroon. It is likely that bush meats have been harvested to unsustainable levels. The animals of choice are monkeys, antelopes, porcupines and grass cutters. However, with the gazettal of the Reserve as a Strict Nature Reserve there has been the loss of a large area of the tribe's former lands for hunting, mainly in the Kupe area (Ebua *et al.* 2011).

We should not leave the description of the local people without mentioning an unusual cultural belief. Mount Kupe is considered to have magical properties. Witches are thought to live there and take humans as slaves to work plantations. There is much more to this story but space does not permit us to discuss it in detail. We know that on a worldwide scale many people have believed in witches and horrible crimes have been committed, for example the Salem trials in the USA. This is not the case at Mount Kupe.

The success or otherwise of the Bakossi Forest Reserve

The first author of this chapter studied the awareness, perceptions, feelings and socio-economic impacts on the Reserve's traditional land users, dwellers and farmers, when its status was changed significantly. This research was undertaken in February 2014. It is through such research that we gain an appreciation of the need to compensate local people when their traditional practices are curtailed. If the declaration of an area as a Strict Nature Reserve is to be welcomed, successful and adhered to, offering those negatively affected an alternative, appropriate lifestyle, in both a material and spiritual manner, is a fundamental and morally necessary offset.

The research was undertaken by distributing a questionnaire to a sample of the Reserve community. Data were gathered on demographics, for example the length of time the respondents had lived in the Reserve, their ages, etc. The major questions related to the respondents' knowledge and awareness of the Reserve, and their understanding of the implications of the forest being protected under very strict rules. The survey also aimed to determine the respondents' perceptions of, and feelings towards, their traditional lands. The people interviewed were asked if there had been a loss of farm lands, restrictions on hunting and overall accessibility to the land, and if there were provisions for compensation.

The relationship between the inhabitants and local forest authorities was also subject to inquiry. Were the inhabitants willing and able to work with the Reserve managers to meet the conservation objectives? As is evident, the questions focused on forest users: those individuals who had historically used the forest land for their livelihood, predominantly for activities such as farming, hunting, fishing, collection of firewood and timber, and gathering traditional medicine, spices and nuts.

Respondents were selected from four villages with a combined estimated population of 15 000, living within 5 km of the Reserve. These were people who would have traditionally used the Reserve. A total of 102 questionnaires were distributed. This sample size was considered adequate given that the population was relatively homogenous. This had been demonstrated by previous surveys undertaken in the area (Ebua et al. 2011). Hatfield et al. (2002) recommended that, in such circumstances, an increase in sample size would not result in a significant change in the overall results. The response rate was 81.4%.

The questionnaires were hand-delivered to respondents (one per family) in their homes and were retrieved later the same day or the next day. This practice, known as the drop-off and pick-up technique, was considered the most appropriate by Olsen et al. (1998) who observed that it ensured a high response rate. Allred and Ross-Davis (2010) also suggested hand-delivering questionnaires, because it provided face-to-face contact with respondents and strengthened their motivation to complete the survey. On the first visit to each respondent, data collectors explained the objectives of the research and, where necessary, interpreted questions. For respondents who were illiterate and consented to being surveyed, the questionnaire was administered verbally by the interviewers.

What were the findings?

Most male and female respondents were aged between 41 and 50 years, and had lived in association with the Bakossi Forest Reserve for more than 10 years. The majority (68.7%) had lived in the area for at least 16 years. Most of the respondents' families had been closely associated with the Reserve since before its protection. Although Bakossi tribal traditional laws regulated the extraction of resources from the Reserve, 83% reported that in this matter they disregarded the traditional law and, before the Strict Nature Reserve's establishment, freely accessed everything they required from the Reserve.

When invited to express their feelings and attitude about the change in status of the Reserve (to a Strict Nature Reserve), most respondents (79.5%) said that they were happy or very happy with their lands being protected in this way. This widespread support for the protection of traditional lands could be traced to the role that the Reserve continues to play in the protection of culture and traditions of the indigenous peoples. If the change had resulted in a major impact on the traditional owners' culture and traditions, we predict that a high number of respondents would have had negative feelings about the change in status, and this would presumably have jeopardised efforts for long-term sustainable management of the area.

Because of a favourable attitude to the change in status, the social impacts were viewed by the indigenous people as minimal. More than half of the respondents indicated that apart from changing or inhibiting their hunting and farming pursuits, the establishment of the Reserve had not significantly influenced their culture and traditions. They indicated that this was because the protection of the lands strengthened the protection of their sacred shrines. This indicates the dominance of religious and cultural matters over material ones. It must be noted that the new status of the Reserve had not introduced restrictions on performance of traditional rights, provided they did not change the landscape or interfere with the composition of the biodiversity of the Reserve.

Social coherence tends to be a major consideration for traditional people such as the Bakossi tribe, and this would have played a role in the broad acceptance of the area's protection, at least during the development phase (NBS 2011). The formal protection of the lands as a Strict Nature Reserve provided long-term protection for the indigenous peoples' sacred areas, and thus was perceived as a positive benefit. This was viewed as compensation (we can call it an offset) for the loss of certain material rights.

However, notwithstanding the social benefits and despite the support for the Reserve's protection, many of the respondents said they would not support an expansion of the boundaries of the Reserve. This and any negative or ambiguous reaction was not based on opposition to the concept of conservation of their lands. Their concern was for additional impact on their material well-being. This is especially an issue in poor countries where there is a lack of alternative sources of livelihood offered to replace the removal of rights to traditional lands.

Based on the respondents' self-assessment of the value of the products derived from the forest area, and the use of local market prices, we understand that a forest-dependent household income in the Bakossi community before protection as a Strict Nature Reserve was considerable. The income from the sale of forest products was much more than the monetary value (about US$2000) of the products consumed by household members. This situation is typical of most rural areas of Cameroon where inhabitants tend to sell almost all the goods they produce to raise money for other needs; for example, for education, health and social activities (Fonchingong 2010).

It is obvious from these estimates that before the strict nature designation, the indigenous families depended heavily on the Reserve for farming saleable products. Only 6% of respondents reported that they had not lost the use of land. Some 82% had lost large portions of their lands to the Strict Nature Reserve, many more than 5 hectares (12 acres). More than half (56.6%) of the respondents reported that their family had been totally removed from their ancestral lands.

Here we come to a crucial point. All of those interviewed said that their families had not been compensated for the loss of the income they had earned from farming their lands, and no alternative livelihood strategy had been presented. The only offset was that protection of sacred sites had resulted, but even that could be diminished if there were not sufficient funding to enforce protection.

Due to the reduction in land to support their livelihood, 71% of respondents reported that there had been a rise in the number of land-related conflicts within the community. The loss of land had encouraged illegal activities within the Reserve. In turn, there have been many negative encounters between forest guards and residents. For example, many respondents (68.7%) reported that they had been chased from the Reserve by forest guards. Of those caught, 72.3% stated that the guards confiscated whatever forest products they had collected. In a related study, Lambi and Ndenecho (2009) reported that 57% of forest guards' time was spent on policing the indigenous people misusing the Reserve, while only 27% of their time was spent on other important activities such as rehabilitation of degraded areas.

The high number of reported encounters between the forest users and forest officers could be interpreted as a strong commitment by the officers to enforce the protection of the Reserve. However, it may also be an indication that indigenous people, although supportive of the protection of their land for religious and cultural reason, have failed to commit to the conservation objectives of the Bakossi Forest Reserve at the expense of their livelihoods. Poaching is undoubtedly driven, at least in part, by the inability of the traditional owners to make an alternative living.

Most of the respondents (85.5%) acknowledged that the Bakossi Forest Reserve was owned by the government; however, many demonstrated limited knowledge of the Reserve as a truly protected area: its precise boundaries, its strict nature protection purposes and its large size. For instance, 34.9% of respondents reported that they did not know whether they lived inside or outside the boundaries of the Reserve. In fact, 45.8% claimed that they lived outside the boundaries although they are most likely to be living within the Reserve. The assertion that they lived outside was possibly a way of avoiding identifying themselves as not knowing the boundaries.

On the assumption that, as reported, approximately one-third of the local people do not know the boundaries, this suggests that the Cameroon government has failed to engage sufficiently with the indigenous users of these lands. This apparent lack of knowledge of the Reserve's boundaries may account for why a substantial number of respondents (41%) indicated that the protection of the Bakossi Forest Reserve has made no difference to their activities. Otherwise, they either do not know the restrictions of the change in protection status, or simply ignore the restrictions that have been imposed on the use of their traditional lands, and continue to undertake what are now illegal activities. Support for this conclusion may be drawn from the comments of the respondents who said they did not use the Reserve although they actually did so. The point is that they misreported their use of the Reserve because they did not want to be identified as violators of government laws.

Some of the illegal use may be because information dissemination on conservation initiatives is primarily the responsibility of the Ministry of Forestry and Wildlife, which has limited financial, material and human resources to engage with the people who live in association with protected areas (Lambi and Ndenecho 2009).

Conclusion

It is widely believed that protection of lands under an appropriate IUCN classification system remains the best strategy to conserve significant landscapes, ecosystems, biodiversity and species (Dudley 2008). Cameroon has embraced this concept. Simply providing such formal protection may stop the short-term and, hopefully, final loss of unique values, but without the funds to provide alternative livelihoods and manage the protected areas beyond their gazettal, detrimental impacts will continue. The land will be degraded, albeit at a slower rate than if it was not protected (Burgin and Zama 2014). In addition, as has been illustrated in this case study, protection may dispossess people of their traditional lands without any compensating provisions; that is, without adequate offsets. The new benefits of protecting sacred sites did not compensate for material losses. As has occurred with the indigenous peoples of Bakossi Forest Reserve, without an alternative livelihood, there is likely to be no option but to become trespassers and poachers on 'their' lands.

This issue is by no means restricted to Bakossi Forest Reserve. Around the world, many protected areas have failed local indigenous people (Redford and Fearn 2007). The problem is not only the lack of finance – although it is crucial – but the question of what is an appropriate offset? While in an industrialised, high-income country, those who lose jobs and annual income when an area is taken for strict protection can be – and inevitably are – compensated in monetary terms for forgone income and, if workers, offered retraining, it is not obvious that this type of offsetting would be realistic and acceptable if applied to tribal folk.

To deal with this issue, governments and other stakeholders must provide alternative sources of income for indigenous people to offset their loss of livelihood. If funding to support those who have lost their livelihood is not forthcoming, as well as enough money to support the sustainable management of the protected areas, it is inevitable that the reserves will suffer environmental decline over time (Dudley 2008). Furthermore, their religious and cultural values, being one positive outcome of preservation as a reserve, could be diminished through lack of resources for maintenance.

Sourcing adequate funding is an enormous challenge for poor countries unless they are given considerable help from rich donors. Even with such support, we are looking at the most difficult of human issues – changing behaviour. This does not mean initiatives to protect extremely valuable environments should cease; to the contrary, protection is necessary if the challenge of sustainable development as formulated by the UN World Commission on the Environment and Development is to succeed.

Clearly, offset policies and practices that are common in affluent countries are not designed to meet the requirements of the poor countries. Addressing this matter is of immediate importance. If not resolved, the planet will eventually be denuded by the desperately poor doing no more than trying to feed themselves.

The role of offsets in the conservation of cultural and built heritage

Johari HN Amar, Lynne Armitage and Tor Hundloe

In this chapter we focus on achieving sustainable development goals – environmental, economic and social – in Tanzania, through the conservation of the country's cultural built heritage. Preserving cultural heritage in developing countries is extremely difficult unless the heritage sites attract large numbers of paying visitors. With small budgets and urgent economic and social problems to be addressed, foreign tourism is the means of providing the funding to acquire, if need be, and maintain built heritage.

Overseas tourists spending currencies such as the euro and US dollar are as valuable as exports. Tanzania gets in the order of 10% of its Gross Domestic Product from foreign tourism and generates ~400 000 jobs in this sector (World Bank 2015). However, with exceptions such as the famous Tanzanian Northern Circuit focused on Serengeti National Park, the location of tourist attractions is important if they are to attract visitors. For example, if the attractions are relatively close to major airports servicing international tourists, they are most likely to be visited. If not nearby, the cost of travel can be a disincentive. If land transport is of poor quality – the journey slow and uncomfortable – this is another disincentive for tourists to travel to a heritage site.

The Northern Circuit tourist sites are advantaged by their relative proximity to Kilimanjaro International Airport. The town of Arusha, with its associated national park, is about half an hour's drive from the airport, allowing a visitor to hire a guide and 4WD vehicle and head the considerable distance to Serengeti. Serengeti gets in the order of 250 000 visitors per year. The World Bank (2015), in its report *Tanzanian's Tourism Futures: Harnessing Natural Assets*, suggests that the Northern Circuit is reaching its carrying capacity for tourism. Too many tourists will damage the wilderness appeal of the area. The World Bank (2015) concludes:

Tanzania has neither fully leveraged its immense endowment of potential tourist attractions nor the opportunities for poverty eradication that the tourism sector offers.

The other part of Tanzania which is a very successful tourist destination is the island archipelago that, prior to the formation of Tanzania, was known as Zanzibar. Tourism is the number one foreign exchange earner for Zanzibar. Tourists can fly direct to Zanzibar International Airport or arrive via ferry from Dar es Salaam on mainland Tanzania. In the order of 175 000 tourists visit Zanzibar per year. The archipelago has many natural attractions, including numerous beautiful beaches, crystal-clear waters plus coral gardens suitable for snorkelling and diving. Then there is its built heritage – of this we will say more.

Initially, we need to explain the dilemma faced in preserving city heritage buildings or precincts.

Challenges in preserving built heritage

There are factors mitigating against the preservation of built historical and cultural sites, and they are not confined to poor countries. Not every city is a Rome or Athens where history is the city's signature. Protecting ancient and, let us not overlook iconic modern, architecture requires those with the power to act wisely. Not all that is new is beautiful! History is lost forever – we do not have a time machine to take us back to the past – when the old is bulldozed for shopping centres and apartment blocks with very short economic lifespans (e.g. 60 years). That rubble on the ground was a centuries-old building – or even older, in the ancient civilisations.

If the buildings that deserve protection happen to be in a relatively poor city where an expanding population causes an ongoing demand for land (or airspace for high-rises) and there is money to be made through a change in land use, there will be temptation to demolish the old and replace it with the new. There is yet a stronger driver favouring demolition. The powerful economic interests of those in the construction industry and real estate sector play a role that poor city governments cannot resist. A historic and cultural area, even one attracting reasonable tourist visitation and consequent spending in the local market, is not going to generate as much income to real estate owners and to governments through land taxes as residential high-rise apartment buildings do. In this context, offsetting schemes have a very important role in protecting heritage buildings in poor countries.

The Tanzanian situation

One of the Tanzanian Development Vision 2025 goals is to have 'a competitive economy capable of producing sustainable growth and shared benefits' (Planning Commission 1999). Yet legislative frameworks like the National Environmental Policies of 1992 and 1997, which outline the environmental policy for Tanzania's mainland and islands respectively, do not include the powers needed to address the threat by demolition, or simply neglect, of heritage buildings. The contribution of built heritage conservation as an element in reaching the country's sustainable development goals is overlooked.

To illustrate the trials and tribulations of built heritage preservation in Tanzania – and to make the point that there is much of significant built heritage value in the country – we offer an overview of its history and geography.

A brief introduction

The United Republic of Tanzania was created with the joining of Tanganyika and Zanzibar. The latter is an archipelago comprising Unguja Island, Pemba Island and several smaller islets in the Indian Ocean. The nation shares its border in the north with Uganda and Kenya, in the west with Burundi, the Democratic Republic of Congo and Rwanda, in the east its coastline borders the Indian Ocean, and Mozambique, Zambia and Malawi lie to the south. Tanzania's population is ~45 million, living in a total area of 945 454 km². The main geographical features include the Great Rift Valley, significant lakes such as Lake Victoria, and Mount Kilimanjaro, the highest mountain in Africa.

Tanzania from which we all came

Tanzania has a very special place in the evolution of the human species, dating back 4 million years when the first human hominid *Zinjanthropus boisei* emerged near the Olduvai Gorge in the Eastern Serengeti (Amar 2017). It is also occupied by the oldest cultural group of Hadzabe or Hadza, the first modern human, *Homo sapiens*. Today, it is the homeland of around 120 native tribes.

The modern built heritage

A unique, rich cultural built heritage developed in Tanzania over the centuries (Hussein and Armitage 2014; MNRT 2014). A standout feature is the traditional settlements of *Inkajijik* houses on the Maasai escarpments. More generally, there are archaeological ruins, historical towns, designed landscapes and gardens, very old cemeteries, a much more recent colonial architecture and an industrial heritage introduced by different occupiers and settlers. There are 44 historical sites and monuments, dating from the 19th and 20th centuries, listed under the *Ancient Monument Preservation Act*. Of these, there are 21 on Unguja Island and 23 on Pemba Island.

On the Tanzanian mainland (previously Tanganyika), there are ~1500 sites that have cultural value, but only 128 are protected by being gazetted, of which 16 are protected under the Tanzanian Antiquities Division. Three heritage sites and Zanzibar Stone Town are inscribed on the UNESCO World Heritage List.

Given the richness and unique aspects of Tanzania's built heritage, legislated regulatory frameworks have been put in place for heritage protection (the most relevant heritage legislation is summarised in Table 11.1). However, as noted above, there are shortcomings – in practice if not in principle. Some of this legislation is a legacy of the colonial past.

The present situation

In spite of the efforts of successive governments and the laws that have been enacted to preserve cultural heritage, very few historic buildings, monuments and sites have progressed from being identified to being recorded and documented, to being protected.

From the 1960s, conservation projects focused on historic buildings, monuments and sites that would help the development of the tourism industry and lead to associated investment opportunities (Ichumbaki 2012). This was a far-sighted policy. However, the practice fell short. The reality was that the long-term economic value of preservation was over-

Table 11.1. Principal legislation for built heritage conservation in Tanzania

Place	Legislation	Objective
United Republic of Tanzania (URT)	Cultural Policy of the URT, 1997	A guide for conservation, development and promotion of cultural heritage that embodies all aspects of Tanzanian lives, including tangible heritage e.g. relics, sites, museum, archives as well as natural and intangible heritage aspects.
Tanzania mainland (Tanganyika)	Antiquities Act No. 10 of 1964	Protection of cultural heritage resources, which are sites, monuments (buildings), relics and any protected objects (e.g. wooden doors).
Tanzania islands (Zanzibar)	Ancient Monument Preservation Decree Cap.102 of 1927	Legislation for the preservation of objects, ancient monuments and historical sites, including any portion of land adjoining a monument site of archaeological, historical or artistic interest on land or in water within the boundary of Zanzibar.

looked in the interests of short-term profits by those who were prepared to demolish the old and build the new. Much of the built heritage was perceived to have no prospect of generating as much money as what we conventionally call, 'real estate development'. By definition this means 'new'. Hence, historic buildings were demolished, or neglected and allowed to gradually deteriorate if demolition was not legally permitted. This permitted the development of a modern built environment at the expense of historic buildings, monuments and sites (Amar 2017).

A digression on values

As an aside, it is important to note that there can be a significant difference between the monetary values that result from market transactions, such as a tourist paying an entry fee to access a historic site and the cost of travel to the site, and what economists call the total economic value of a good or service that has considerable non-market value. These latter values are external benefits that cannot be appropriated by the owner of the heritage building or site. They can be of considerable monetary value to others in a country. But only if the market-derived income from preservation is large enough will an owner have an incentive to preserve the heritage property.

The non-market values go by various names. One is existence value, whereby those who do not visit the heritage site, and are unlikely to, nevertheless value the fact it exists. Numerous economic studies have found that people, if rich enough, are willing to contribute to the preservation of heritage properties. There is also option value, which is likewise a willingness to contribute to the preservation of a heritage property. In this case the person making the contribution wants it to be preserved as they might want to visit it in the future. And there is bequest value (alternatively called inheritance value), whereby a monetary contribution is made to protect the heritage property so that future generations can visit it, or simply appreciate the fact that it exists.

All this is fine, except the existence of these non-monetary benefits does nothing to help a poor nation obtain the funds it needs to purchase and maintain heritage properties. In theory, an international conservation organisation could seek and collect donations from relatively rich people and funnel that money to the government of a poor country for the explicit purpose of preserving a heritage building or buildings. That aside, what can be done – and is being done in several instances across the globe – is to estimate, or guesstimate, what sum of money can be gathered from foreign tourists, over and above their necessary expenditure, and add it on as a special fee charged to enter and view famous sites. The total amount spent by visitors might be enough to save a historic building or site from demolition. It might fall short of that monetary target, but it is a practical step. Foreign visitors who are likely to spend thousands of dollars on airfares and accommodation find reasonable entry fees no burden. Across the globe, this approach is common for entry to famous national parks, and there is no reason to not apply it widely in Tanzania.

Demand for land and residential accommodation in a market economy

Currently, Tanzania's population, economy and urban development are growing rapidly, leading to strong demand for a range of infrastructure to be built. Geography is important in Tanzania. The country's population is sparsely distributed across the regions, with an overall annual growth rate of 1.96% but reaching 4.77% in urbanised areas. This urban growth is driven by migration from the rural areas as well as the birth rate. The consequent rapid urbanisation has resulted in a considerable amount of Tanzania's built heritage being threatened, left to deteriorate or demolished at an alarming rate (Ichumbaki 2012). We have already addressed some of the incentives leading to this situation. There are additional ones which we introduce now.

One of these factors is a limited understanding of the importance – and potential for foreign exchange earnings – of the nation's historical buildings and cultural sites. However, it needs to be recognised that it is much to ask of a poor country to wait for tourists to arrive in sufficient numbers to justify preserving its cultural environment. A poor country cannot do as the French (Paris), British (London), Italians (Rome) or Greeks (Athens) do – preserve the old and pay for it from the income generated by their very large, cultural tourism sector. These nations and their cities have established a reputation – the marketing people would say a 'brand' – and need do no more than await the paying visitors. There are heritage precincts in Tanzania where this model applies – but there is also a difference between the economics of, and attitudes to, preservation in Dar es Salaam and Stone Town.

Where rapid urbanisation is a major reason for the demolition of heritage buildings, there is an immediate imperative to formulate a means whereby the immediate needs of urban housing and infrastructure can be met while preserving the valuable old buildings and sites. Preservation is a pro-environment action and, in the Tanzanian case, a policy has to be developed which offsets the financial loss caused to the owner by maintaining the old when income could be earned by replacing it.

A little history

It is helpful at this stage to present a historical overview of Tanzania's planning, development and economic policies. The discussion revolves around three policy reforms introduced during the post-independence period in Tanzania (1960s–1990s): urban planning initiatives, transferable development rights, and a free market economy. The significance of these three reforms is that they occurred at a time when both the Tanzanian mainland and islands were experiencing deterioration of their urban built environment and failing infrastructure. This was due to a lack government funding, exacerbated by the large number of arrivals in the cities, particularly Dar es Salaam, leading to overcrowding, widespread disease and urban poverty (Sheriff 1995; Nnkya 2007; Amar 2017).

During the 19th and 20th centuries, both Tanganyika and Zanzibar experienced periods of significant growth in the urban built environment. Very different forms of building, monuments and cultural sites, reflecting local culture as well as that of the German and British colonial powers, were built. These buildings and places are today referred to in law and policies as 'built heritage', and have long been acknowledged in town and country planning ordinances, as Tanganyika's *Town and Country Planning Ordinance 1956* and Zanzibar's *Town and Country Planning Ordinance 1955*. These provide a legal framework for comprehensive planning schemes that, among other necessary things, focus on rectifying environmental problems and dealing with economic circumstances and poor living conditions. The ordinances have been used as legal instruments to create and execute various master plans, along with conservation plans for cultural built heritage (Lall *et al.* 2009).

An early example that illustrates this is the German *Building Ordinance 1891*, which was later adopted by the British regime as the *Town Development Control Ordinance 1936*. This ordinance divided Dar es Salaam into three distinctive spatial zones, subdivided by ethnicity and economic class (Armstrong 1986). The conditions for occupation of the various localities were (Brennan *et al.* 2007):

- Zone I: required residents to construct sturdy European-style buildings;
- Zone II: required residences to be 'other' building types but not negerhutteen ('negro huts');
- Zone III: no construction requirements were required. However, the *Township Rules 1923* required construction of Negro huts built with mud and a thatched roof.

The 1949 Dar es Salaam Master Plan was designed to implement this development pattern (Nnkya 2007). In concert with the *Antiquities Act 1964* Conservation Plan, the planning scheme involved the protection of the socio-economic and environmental historical significance of Zones I and II. Today these form part of Dar es Salaam colonial town (Hoyle 2002). Under the 1968 Master Plan, the built environment in Zone III was redeveloped to tackle the failing infrastructure and associated overcrowding, disease, environmental pollution and poverty (Armstrong 1986; Nnkya 2007).

At the same time, Zanzibar's *Town and Country Planning Ordinance 1955* enabled the relevant government minister and planning authority to manage, supervise and if necessary enforce land development in Zanzibar (Haji *et al.* 2006). The ordinance gave additional powers to regulate and manage all matters related to the conservation of heritage properties. Under this ordinance, the Kendell Plan of 1958 and the German Master Plan of 1964 (Siravo 1996) were drafted to tackle urban problems similar to those in Dar es Salaam.

As made clear by Amar (2017) and Haji *et al.* (2006), the main objective of the two Master Plans was the provision of high-rise buildings on the outskirts of urban areas, where the growing urban population settled. This policy had the potential to revitalise inner-city historic areas by reducing traffic congestion, easing population pressures and relocating industrial activities closer to the new population centres.

The most controversial issue in the Master Plans is that the planning schemes and development initiatives contained therein did not require a Conservation Plan that mandated assessment of the country's historic built environment. For instance, the 1968 Dar es Salaam Master Plan ignored the conservation of colonial infrastructure and the built environment (Kasala 2015), as its priority was government development schemes driven by market forces (Masele 2012). There was no room for non-market values. Kasala (2015) notes that when the 1979 Master Plan was implemented, its long-term planning and development schemes facilitated the adaptation of the international style of high-rise, modern buildings; no deliberate efforts were directed at heritage protection. This emphasis led to the replacement of privately owned historic buildings, monuments and sites with modern buildings (Lwoga 2013). While the situation was different in Zanzibar, where Master Plans aimed to ensure that its historic town sites remained intact, some historic buildings, monuments and sites were allowed to decay and collapse during the 1970s (Sheriff 1995).

The overriding problem for the country was the lack of a heritage offsets scheme which would provide incentive structures to favour preservation of heritage values, while permitting urban development for the increasing population. Serious consideration of various approaches to heritage protection was warranted (Siravo 1996; Haji *et al.* 2006). Clearly, as observed by Elefante (2007), without a formal, dedicated Conservation Plan or a heritage offsetting system, the protection and management of cultural built heritage is compromised.

Tanzania's Master Plans, along with town planning and development reforms, attempted to eliminate the 'racially' segregated zones in Dar es Salaam, and remedy the economic, environmental and social problems that arose when a large population of squatters arrived with large-scale migration to the cities in the 1960s (Brennan *et al.* 2007).

Komu (2011) states that in this era around 90% of the 790 000 people living in Dar es Salaam were squatting; this was just above 6% of the nation's population.

The legacy of history

To allow for a more complete understanding of the reform of the urban situation – successful or not – a little more historical knowledge is required. We have mentioned German and British laws without explaining how and why they came about, and what independence meant.

We tend to know more about Zanzibar than the mainland due to its location in the Indian Ocean and the ability of sailors to use the seasonable winds to sail south from India, Persia (modern Iran), Oman and Yemen and return when the winds favoured the northward journey. It is possible that sailors arrived as early as the first century AD. Certainly, by the 11th century there were people from the Arabian Peninsula living in Zanzibar, as evidenced by a mosque of that period.

Where we are on concrete ground is the recorded voyage of Vasco de Gama who arrived at Zanzibar in 1499. Following his visit, the Portuguese had a presence there for ~200 years. Then in 1698, the Sultan of Oman took control of Zanzibar and the east African slave trade began. An Arab elite took the land while the indigenous Bantu people were reduced to serfs. Slaves were brought in from the mainland of east Africa. Those who were not forced to work in the clove plantations were sold to Arabia, Egypt and Persia. The extremely cruel and degrading treatment of the slaves left a legacy of hate which exploded in 1963 with the Zanzibar Revolution and the slaughter of tens of thousands of Arabs and Indians. Some historians consider this a truly proletarian revolution based on class. Arabs were the dominant land-owning and ruling class while Indians owned most businesses. Other analysts place more store on ethnic differences. A combination of the two is a likely catalyst for the revenge which took place.

Much happened between the 17th-century Omani takeover and the revolution and independence of 1963. In 1840, the Omani Sultan was so certain that he owned Zanzibar that he made Stone Town his capital. This situation did not last, and by 1890 Zanzibar had become a British protectorate. Britain had formally abolished slavery some decades earlier and once in control of Zanzibar it rid it of slavery.

While these events where fashioning the society, economy, culture – and our main focus – Zanzibar's built heritage, mainland Tanganyika was going through its own trials and tribulations as a colony or quasi-colony of European powers. Until the end of World War I, it was part of German East Africa. With Germany's defeat, it became a League of Nations mandated territory controlled by the British. Following World War II, Tanganyika became a UN trusteeship of Britain. The architecture in its capital, Dar es Salaam, reflected the various colonial influences.

In both Zanzibar and Tanganyika foreign religions (Islam in the former, Christianity in the latter) took hold with their imprint on houses of worship. And of course the African culture remained – more appropriately, the African multitude of cultures and beliefs.

This history has led to an eclectic, fascinating built heritage. In Stone Town, much of it remains. There is no better place than here to explain the name of this town – the stone used in building was actually coral, of which there was plenty in the surrounding waters.

Independence, socialism then a free market economy

On 9 December 1961, Tanganyika became an independent nation, led by the most revered African leader, Julius Nyerere. Independence was achieved without bloodshed. In 1963,

Zanzibar was granted independence by Britain and a year after the revolution in 1963 it joined with Tanganyika to become the United Republic of Tanzania.

Julius Nyerere was a Christian socialist who had a major influence throughout much of Africa in its post-colonial days. African, or more specifically Tanzanian, socialism was codified in the Arusha Declaration of 1967. Space does not permit the full story of the achievements and failures of Tanzanian socialism. There were major advances in education; however, the Tanzanian version of collective farms did not suit the tribal farmers despite Nyerere's personal view that African farming had been a collective pursuit. Overall, Nyerere's major achievement was bringing a sense of pride to his people. Recognised as a philosopher and a humble human being, he achieved what few other African leaders have done. He is still revered.

The new country was a proud, socialist state bent on self-reliance, despite both Cold War adversaries' provision of money and expertise. Nyerere sought to create equality among citizens. He also sought to distance the new nation from the colonial overlordship it had experienced for a long time. Discrimination in zoning of the built environment, discussed previously, was abolished.

The government of the United Republic of Tanzania legislated to acquire and declare buildings owned by Europeans, Arabs and Indians in the colonial towns as government properties (Sheriff 1995; Ichumbaki 2012). The historic buildings, monuments and sites thus acquired were entrusted to the National House Corporation and to the *Waqf* and Trust Centre under Tanganyika's Arusha Declaration of 1967 and Zanzibar's Presidential Decree No. 13 of 1965, respectively. Under the declaration and decree, the two institutions were responsible for conservation, management and development of the unique cultural built heritage on behalf of the government (Cunningham Bissell 2007; Ichumbaki 2012).

Khalfan and Ogura (2012) define *waqf* as 'an endowment of a property carried out voluntarily by individuals as everlasting charity, which once endowed, is subject to religious rules' where the rules state the property is not to be sold, mortgaged, gifted or inherited. Its management system is guided by an investment model in which properties are leased to generate profits, of which part is reverted to conservation in perpetuity and part for the improvement of the socio-economic well-being of the beneficiaries and humankind in general (Oberauer 2008). Khalfan (2014) reports that *waqf* minimised the demolition by neglect of historic buildings in Tanzania, reducing it to around 3% in 1993, compared to the loss of 7% of privately owned properties in the same period.

As a result of this conservation practice, over 600 *waqf* historic buildings representing diverse cultures and architectures contributed to Stone Town being inscribed in the UNESCO World Heritage List in 2000 (Sheriff 2001; Yahya 2013). This indicates that *waqf* is an extremely effective tool. It is not dissimilar to the National Trust in Australia and the UK, and national trusts in general.

Through its representative institutions, the government took the bold step of removing Stone Town from the conventional real estate market. By doing that it achieved the kudos of becoming a World Heritage site, and consequently a drawing power for tourism. We can anticipate that the long-term economic gain from tourism will offset the forgone income from demolition and conversion to a modern, but very ordinary, inner-city environment.

At the time of independence, the National House Corporation was tasked with providing affordable housing for Tanzania's increasing population, but was left to its own devices by the central government of Tanganyika due to limited financial resources. Its initiatives were to keep existing urban colonial buildings as heritage (Brennan *et al.* 2007) and to construct affordable houses through issuing mortgages and directly purchasing existing houses (Owens 2012). Approximately 6000 houses were acquired between 1972 and 1984 under the *Acquisition of Buildings Act 1971* (Owens 2012).

Here is an appropriate place to digress once again and discuss the economic situation in Tanzania in the 1970s.

Drought, war, oil shock and the IMF

Tanzania experienced moderate economic growth until a major drought hit in 1974–75. With a return to normal weather patterns, crop yields increased and annual growth was in the order of 6% – not unreasonable for a very poor country with little industry except agriculture. However, in 1978–79, two disasters threw the economy into deep trouble.

Idi Amin, the Ugandan dictator, had an antagonistic approach to its neighbour Tanzania from the early 1970s, mainly because Tanzania gave shelter to Idi Amin's opponents – a considerable number lived in Tanzania. In November 1978, Idi Amin's forces, assisted by thousands of Libyan troops, invaded Tanzania. The Tanzanians won the short war, but there were drawn-out adverse effects on its economy.

Nyerere became far more conscious of the possibility of future invasions. Approximately 48 000 people were resettled from northern Tanzania, at significant cost. Many had seen their homes burnt or ransacked, their livestock slaughtered and their livelihood destroyed. There was a further very large cost involved in boosting the country's defence forces and preparedness. The impact on Tanzania's fragile economy was devastating – money which could have otherwise gone into building social infrastructure, including much-needed urban housing, was no longer available for that purpose.

This undermined Nyerere's socialist dream for Tanzania, but the 1978–79 oil shock – a very substantial increase in oil prices decided by a cartel of Middle East oil producers – dealt it a near-fatal blow. Tanzania had to approach the International Monetary Fund for assistance. It was given, but with the proviso that a free market economy replace the current mildly socialist system. This reform had a major negative impact on the country's financial ability to preserve heritage buildings, especially in Dar es Salaam where there was no equivalent to the Stone Town *waqf*.

Economic challenges to preservation of heritage buildings

The National House Corporation was forced to rent the majority of colonial heritage buildings to the local population, in order to shift the repair and maintenance costs without altering the land use and development patterns enforced in the Master Plans. This, along with the *Rent Restriction Act 1962* which set a rent ceiling of 10–12.5% of 75% of tenants' gross income, led to a reduction in new construction.

However, the condition of the heritage buildings deteriorated because tenants believed that the government owners should cover the costs of maintenance, rehabilitation and restoration (Amar 2017). This did not occur in Zanzibar as the *Waqf* and Trust Centre, instead of using tenants' rents to cover such costs, allocated the money to the maintenance of religious buildings, salaries for imams and the support of religious education, as well as other non-heritage protection projects (Sheriff 1995). Similarly, the National House Corporation used any surplus to construct new buildings – some residential, others commercial and industrial – all of which were more profitable in a competitive market than in renting historic buildings subject to rent control.

As a consequence of the free market reforms, occupants in both Tanganyika and Zanzibar abandoned historic buildings. There were three subsidiary causes: high maintenance costs, the introduction of mortgage finance which allowed the purchase of a dwelling over many decades and a developing preference for modern architectural styles and associated new technologies (Sheriff 1995; Cunningham Bissell 2007). In much the same way that tenants moved out of heritage buildings, private owners, developers and investors had little or no interest in this sector of the real estate market. This neglect was not confined to the large cities; historic dwellings, monuments and sites in other parts of the nation suffered the same fate. Many collapsed or were demolished due to decay (Sheriff 1995; Cunningham Bissell 2007; Ichumbaki, 2012).

There was another factor that led to a lack of interest in built heritage preservation. During the 1980s, nature-based tourism was becoming more important as a driver of Tanzania's future economic prospects.

Nature-based tourism vs cultural tourism

The tourism industry became a priority (Amar 2017). It was not just about money. Tanzania has a fascinating pre-human history and some of the most spectacular wildlife in the world. These are things of which the nation is very proud – knowing that people want to visit, to learn of pre-history and engage with wildlife, was uplifting although most citizens would gain little if any economic value from tourist operations. This, of course, is not peculiar to Tanzania. Trickle-down economics is at best an English mist not a tropical downpour.

Cultural tourism was a potential drawcard and money-earner for the Tanzanian economy which, if pursued, should have led to preservation of historical buildings and precincts. However, it would have required promotion of colonial architecture, both ancient from the days of the Indian and Arab colonialists to the much more recent German and British buildings – and the mainland of Tanzania did not want to focus on its colonial past.

Zanzibar, however, took a different approach. In the 1980s, planners in Zanzibar were of the view that a Master Plan that promoted cultural tourism was the best approach to address social, economic and environmental issues, while achieving the conservation of cultural built heritage (Goldman 2014). In 1985, the Revolutionary Council approved a Master Plan (Amar 2017). It focused on removing administrative functions and manufacturing industries from Stone Town, developing three bazaars to provide for shopping and adapting historic buildings into high-class tourist facilities (Siravo 1996).

Significant parts of Zanzibar's built heritage were gazetted and placed under the care of the Ministry of Sports and Culture. These places have the ability to attract revenue that is used to finance conservation (Marks 1996; Hitchcock 2002). They included Beit El Ajaib (House of Wonders) built in 1883 as a ceremonial place, the Palace Museum built in the 1890s to house the Sultanate family (Fig. 11.1) and the Old Fort built in the 15th century. Entrance fees are collected, donations accepted and souvenirs sold – all normal practices.

On the Tanzanian mainland, new perceptions developed about the conservation of colonial urban heritage, especially in Dar es Salaam. This was due to a recognition of its ability to attract visitors and generate much-needed foreign exchange (Lwoga 2013). In conjunction with the 1996 Tourism Master Plan and the *Antiquities Act 1964*, new heritage conservation plans addressed the urban environment and the economic and social constraints that were not dealt with in the 1975 Master Plan. The latter had focused on meeting the needs of the fast-growing urban population (Amar 2017). The new conservation plans categorised Dar es Salaam's built heritage based on historical, aesthetic and architectural values as well as environmental quality. Amar (2017) describes the three main categories of cultural and historical architectural value:

Fig. 11.1.   Historical buildings in Zanzibar. (a) Palace Museum, built in the 1890s. (b) Beit El Ajaib, also known as the House of Wonders, built in 1883. Photos: Johari Amar.

- Category I: buildings of Arab and German origins before 1961;
- Category II: buildings of German, Indian and British origins built after 1961;
- Category III: buildings of over three-storeys with features that do not conform to Categories I and II.

Nature-based tourism

In spite of Tanzania's hopes to conserve its built heritage through cultural tourism and foreign investment, the gap between the goals of cultural tourism and built heritage conservation widened in the 1980s and 1990s. Marks (1996) and Spenceley (2012) state that this was due to far more focus by tourism operators and government institutions on promoting nature-based tourism. Where cultural tourism was promoted it was in the form of traditional dance, art and craft, not the cultural built heritage found in towns. Given the success of tourism based on Tanzania's world-renowned wildlife, the attention to nature-based tourism is not surprising.

Since few historic buildings, monuments and sites were gazetted for protection, there was limited scope to increase tourism in the built heritage sector, except in Stone Town (Amar 2017). There was considerable loss of heritage buildings elsewhere in Zanzibar. Khalfan and Ogura (2010) summarised the technical surveys carried out for development of the Zanzibar Historic Cities Programme. These were undertaken with the assistance of international organisations – UNESCO, UN Habitat, UNDP, SIDA and the Aga Khan Trust for Culture. The surveys discovered that around 55 buildings had collapsed, and a further 60 were in danger of demolition due to neglect.

Lwoga (2013) and Spenceley (2012) noted that, on the Tanzanian mainland, the strategies for tourism development focused on upgrading infrastructure (roads and airports), the service industry, natural heritage conservation (national parks) and poverty reduction – not on the built heritage.

The future of Tanzania's cultural built heritage is uncertain, due to the increasing dilemma of conflicting goals: preserving the authenticity and integrity of Tanzania's significant built heritage values or promoting the modernisation of cities, powered by the commercial property market.

Tanzania faces another dilemma – conservation plans for the management of historic buildings, monuments and sites prepared by heritage departments are often overruled by the ministers in charge of town planning development (Amar 2017). With the exception of Stone Town, very few historic buildings, monuments and sites have been identified, recorded and documented, and thus officially protected.

Building owners, investors and developers eye heritage buildings located in commercial centres, for example, Zanzibar's Stone Town and the Dar es Salaam city centre. These locations are in high demand for luxurious office and retail buildings. New buildings will have modern amenities, including elevators, cooling systems, standby generators, fire detection and parking garages. These add to building costs and inflate and stimulate the property market (Kongela 2013). The gap between high-end office buildings and low-end (heritage) buildings was measured some years ago. The difference was stark, ranging from US$16–20/m² at the high end to US$5–7.5/m² at the low end. As such, the inflation of rents and land prices discourages the conservation of cultural built heritage. It even influences the government bodies that are responsible for heritage buildings. For instance, the National House Corporation petitioned for its historic buildings listed under Conservation Area Notice No. 2006 to be removed from the Antiquities Heritage Register because the organisation was losing money on its investments (Rhodes 2014). The government waived the conservation restrictions on at least 100 significant buildings in 2007 (Amar 2017) in order to pave the way for the construction of modern high-rise buildings that were expected to boost the country's economy. As noted by Rhodes (2014), by 2013 this had led to significant demolition by neglect.

Community stakeholders who seek preservation of heritage buildings but who do not own them or otherwise have any contractual rights in the buildings are faced with the challenge of how to raise funds to finance restoration and rehabilitation projects. They are forced to rely on international agencies to provide financial support. Financial assistance provided by international organisation is a means to the desired end.

The post-independence planning policies implemented in Tanzania narrowed the social and economic gap between town and rural landscapes. The policies pertaining to urban land reforms were well meant. However, the heritage sector in Tanzania has experienced difficulty in ensuring its plans for the protection of cultural built heritage remain intact due to the problem of financing conservation plans. As James Carville, strategist for Bill Clinton during his successful campaign for US President in 1992, quipped, it's 'the economy, stupid'. Retaining and maintaining heritage buildings as residential properties means lower rental income than can be gained from high-rise apartments. This difference in income has to be met, in other words offset. The finance to do so has to come from either the nation's tax revenue or as a financial gift from international donors. This could be only a short-term problem. In the longer term, Tanzania hopes to be able to draw visitors who spend money on local products while they view and reflect on their interesting – even fascinating – surrounds, as in the case of Stone Town heritage precincts.

Doing what was done for Stone Town takes political courage. Tourists might not come in sufficient numbers to justify the cost and there may not be political support for this type of project if there is only a trickle of financial benefits to the average citizen. The *waqf* management system applied in Stone Town is the notable exception, based on a principle of formal heritage protection as with national trusts in rich countries.

Visitors will pay to visit and explore built heritage sites and individual buildings. Money earned in that way allows heritage precincts to come alive. Not every built heritage site is a replica of Venice, but on a much smaller scale any interesting site can work towards that ideal. Success is in no small way the result of tourist expenditure. If those with rights to destroy old buildings believe they can make more money in modern real estate developments, a means of compensating them for their forgone income has to be found. Contrary to the practice in biodiversity offsetting, we cannot ask these people to find an equally valuable site. If the solution is that part of the tourist charges are used for compensation, that is an offset worth providing. Something along these lines has saved Stone Town.

Sub-Sahara Africa is best known for its magnificent natural environments. Its traditional and colonial cultures are far less known. The fact that we all came out of Africa grabs our attention – and more and more of us visit in search of *Homo sapiens'* roots. Undeniably, the planning of heritage offsets is based on the perceptions of stakeholders whose values often vary with individual interest, experience and knowledge. It is important to consider that while preserving colonial heritage can be a reminder of a colonial past, it can also can play a positive role in developing human understanding. We need to learn from history in order to make progress towards a better world.

Conclusion

That the slave market has been recreated in Zanzibar for visitors to view with horror is a strident reminder – if we need one – that never again should this occur, and that human progress is possible. No other country in the world is better suited to tell this story than Tanzania. It is where we can commence to trace our common history on a journey that, we hope, has a long and ever better future.

ALL FORMS OF OFFSETTING ROLLED INTO ONE

The Adani coal mine: attempting the impossible

Tor Hundloe and Sheron Chand

> **A Child's Ode to the Finch**
> *I wandered the red bull-dust track,*
> *In the footsteps of Banjo's outback.*
> *He missed the bird song,*
> *His eyes on the billabong.*
> *For me, the finch I came to see.*
> *The bird had to be somewhere,*
> *I looked here, there, everywhere*
> *I found it resting in a sad gum tree.*
> *I'll come again tomorrow,*
> *In hope, postpone sorrow.*

[Anon. undated b]

In outback western Queensland, in Banjo Paterson country, live a few hundred small, attractive – some would say 'well dressed' – birds, the black-throated finch (southern) *Poephila cincta cincta* – (see Fig. 12.1). This species is classified as endangered. It is very likely that these black-throated finches will be forced to find a new home in order for a large coal mine to be opened. The Adani Coal Mine, named after the mining company the Adani Group, is possibly the currently most controversial project in Australian politics.

We have chosen this project to illustrate the trials and tribulations in environmental offsetting. Four major types of offsetting are involved: biodiversity offsetting, carbon offsetting, offsetting the depletion of a non-renewable resource (intergenerational offsetting) and, if a new home is found for the birds on productive grazing land, as is likely, and that land requires a form of protected arrangement, someone will need to offset the loss of income to the farmer.

Many projects, such as a new dam, airport extension, tourism complex or large mine, result in residual damage, notwithstanding all reasonable efforts to avoid or mitigate environmental damage. There is unavoidable damage if the project goes ahead. When the project is very large, with divergent stakeholder views, determining what are reasonable offsets for the residual damage caused by the development may be extremely complex and protracted. Ultimately, the assessment of the proposed offset should be based on science; however, it is necessary to convince the interested parties (including the public) that the right decision has been made. That in itself can be a difficult task!

Fig. 12.1. Endangered black-throated finch (southern), *Poephila cincta cincta*. Photo: Eric Vanderduys.

A small bird's home in a coal field: the background

The controversy is not simply about finding a new, suitable home for a small bird that few people have ever seen, or likely to see. However, let us commence with its story. The finch lives on seeds that grow in a wide area of Australian grasslands, from northern New South Wales to the far north of Queensland. Its range has been dramatically reduced in recent decades, as a result of cattle grazing and large-scale cultivation, and its habitat and feeding areas have been degraded by feral animals such as pigs.

The Adani mine is to be sited in the Galilee Basin (further discussed below). The Galilee Basin finches are found throughout the Basin, not only on the Adani mine site but at six other mine sites in the Basin. The birds' major home is the Adani mine site.

Today, the finch's still sizable, sustainable populations live near the north Queensland city of Townsville and in the outback area in the Galilee Basin. This basin is considered one of the largest coal fields in the world. It covers 24 700 km², about the size of the UK. If all the coal in the Basin was mined and exported it would double Australia's coal exports.

It is no surprise that there is great interest in mining the coal, and equally unsurprising that doing so is controversial. If all the coal so far identified in the Basin was mined and burned to produce electricity, an estimated 705 million tonnes of carbon dioxide would be released annually – 1.3 times Australia's current annual emissions (Steffen 2015). The existing quantity of carbon dioxide in the atmosphere is already more than twice the amount that was present just before the commencement of the Industrial Revolution in the late 18th century.

For some decades the scientific community, particularly climatologists, have been drawing attention to the generally negative impacts of carbon dioxide on both the human-made and natural environments. Their models have been refined over time and their predictions have become more precise. Yet, a great deal still needs to be better understood for the task of making predictions. The difficulties are not only related to the role of nature (for example, the importance of the oceans as a carbon sink), but guessing (it cannot be more precise) as to how we, individual humans and collectively as nations, will react to the threats of climate change. For many years the world's governments have struggled with the matter and achieved little real success. On the other hand, individuals have installed rooftop solar panels, reducing their electricity bills while reducing carbon dioxide emissions.

Emissions from coal-fired power plants are the major source of greenhouse gas. For this reason alone, coal as a fuel has a limited lifespan. When might the coal age end? If a cost-effective means of sequestering carbon dioxide is developed, it might not end until all the coal is burned. However, at present the prospects of inexpensive sequestering are remote. There is the possibility, increasingly becoming a high probability, that a range of renewable energy sources will replace thermal coal over the next two to three decades. The continually reducing costs of solar and wind power supplemented with an expansion of pumped-storage hydropower will drive this change of fuels. Coking coal is another matter, but it also is to be challenged sometime in the future.

The coal deposits in the Galilee Basin are held by various private companies under leases granted by the Queensland government. An insight into the processes involved in gaining the rights to mine would be a truly fascinating topic, but this is not the place to investigate the peculiarities of mining law in Australia. However, two points need to be made. First, the coal is owned by the Queensland government on behalf of the people. The government could, if it wished, mine the coal by contracting mining companies to do the work, or the government could sell a share (e.g. 49%, as Norway does with its oil resources) but remain in control of the development of the mine. Governments that do not do either of these things face problems in getting a fair economic return for the once-off activity of mining.

The second point follows from the first: the government is expected to do the best deal possible for the owners of the coal, the citizens. Given its unwillingness to take either of the options mentioned above, the only means of gaining some benefit from exploiting the coal is through royalty payments and taxes on profits from the de facto owners of the coal. We should note that any subsidies paid to the mining companies should be subtracted from this government income. If we attempt to calculate the net financial situation (the amount citizens get), we discover that lack of transparency renders this task near-impossible for all but a forensic accountant or economist.

The economic challenge for a de facto owner of a coal resource which is used to generate electricity is to determine the rate of coal extraction. A range of economic factors come into play, including the price of coal today and the expected price in the future. The same temporal consideration has to be given to exchange rates, but in the present era the crucial question for a miner is estimating or guessing the date at which coal will no longer be needed and it will become a stranded asset, and worthless. This will occur when natural gas plus renewable energy sources are capable of meeting total demand. Sometime after that, natural gas will also be unsold as renewables will carry the total load.

Those with coal mine leases are watching the rapid growth in renewable sources for electricity generation and attempting to figure out their future. Do they commence mining immediately and, if so, will an immediate oversupply occur if all do the same thing? Do they mine at a slower rate, knowing that there will be considerable coal left in the ground when the mines are closed? Governments reliant on taxes and royalties from coal mining face the same questions. While they appreciate the income flows from mining coal, they face a degree of public opposition to coal mining. This is clearly the case for the Adani coal mine.

Types of offsetting for the Adani coal mine

One type of offsetting is required to deal with direct damage from the Adani project. This is offsetting the destruction, or when not complete the degradation, of habitat and consequentially the serious impact on endangered animals and plants. The Adani coal mine is intended to involve both open-cut and underground mining. For open-cut mining, all vegetation and massive amounts of topsoil and earth (overburden) are removed, and even greater amounts of earth are dug out in the process of getting to the coal. While underground mining leaves the surface intact, subsidence results and the landscape is altered with flow-on negative impacts. In terms of open-cut mining, plants and animals that are dependent on the mined area are lost – unless safely moved and re-established elsewhere, assuming that is feasible.

Burning the coal once it is mined and transported to a power station or furnace is another dimension to the story, and it requires a different type of offsetting. The combustion of coal releases greenhouse gases. We are at a critical stage, both internationally and locally, of being required to meet commitments to reduce greenhouse gases. Any additional greenhouse gases emitted makes meeting a climate change threshold more difficult. Can the extra greenhouse gas be offset? Who is responsible for this, the miner or those burning the coal?

The third element of the story pertains to the coal itself. Coal is, for all intents and purposes, a finite resource – burned and it is gone. We have relied on coal for over 100 years to power our industries, to illuminate our cities and provide our household electricity. How are future generations going to enjoy a lifestyle similar to ours, without coal? The question becomes, how do we offset the loss of a non-renewable resource? The type of offsetting required in this instance is best addressed by establishing a sovereign wealth fund in which the profits of mining are invested. The idea is to guarantee a constant income flow; that is, the sovereign wealth funds are invested with the aim of generating the same flow of net income as mining did.

This case study discusses four different types of potential losses to be offset: loss of endangered animals and plants, further loss of a clean and appropriately tuned atmosphere, loss of a traditional source of electricity and loss of some grazing land which will require reduced grazing elsewhere to provide a new home for the finches.

Setting the scene: a large coal mine

The Adani coal mine and associated rail line project is officially known as the Carmichael Mine and Rail Project. The company involved is Adani Mining Pty Ltd, a subsidiary of the Adani Group, an Indian company. The market capitalisation of the Adani Group has been reported at US$30 billion. It is a major player in the energy chain in India. If the mine proceeds as planned, it will entail six open-cut and five underground mines. The coal will be transported from the mine by a purpose-built railway to a Queensland port fronting the Great Barrier Reef World Heritage Area, and from there shipped to India. The operation of the port and shipment through Great Barrier Reef waters are themselves controversial matters, but are not dealt with in this chapter.

In making the case for the mine, the Adani company asserted that the project was much-needed because it would bring electricity to poverty-stricken, Indian slum-dwellers, of whom a massive 240 million have no electricity other than that which they can obtain by connecting wires to overhead power-lines. We won't be diverted by arguing the pros and cons of this attempt to justify the mine; however, we need to demonstrate that using coal to bring electricity to the poor is not the only option. It is probably the most costly. How do we make sense of that?

We note an announcement on 29 June 2017 by the World Bank, titled *Solar Powers India's Clean Energy Revolution*, which reported that 'India is emerging as a front runner in the global fight against climate change.' Does India need more coal-fired electricity? At present, one-sixth of India's electricity is provided by renewables and, according to the World Bank announcement, half of the country's electricity will be provided by renewables by 2027. Village-scale, solar electricity based on photovoltaic rooftop panels, batteries for storage, micro hydro-electricity schemes plus, in appropriate places, wind turbines are less expensive solutions to meet the challenge of bringing power to the poor.

Notwithstanding the rather rapid development of renewable energy in India, the demand for coal will continue to grow up to 2027 as a consequence of population increases and economic growth. But the World Bank argues that the increase in demand for coal-based electricity will be in the order of 25–30% while over the same period the growth in solar-generated electricity will be ~250%. This is due to the fact that solar-generated electricity has become as affordable, if not more so, as electricity generated by fossil fuels. It will pay to be mindful of this World Bank assessment when considering the prospects for the Adani mine. By the time the company is ready to send coal to India, it might be too late in the context of the costs of renewables.

The Adani mine is not the only large coal mine proposed for the Galilee Basin. Of the proposed mines for the locality, one is under the partial control of Gina Rinehart and another is associated with Clive Palmer. These Australian household names are a guarantee that there will be considerable public interest in what happens, or is proposed to happen, in the Galilee Basin. Gina Rinehart is thought to be the richest woman in the world, based on the extensive mineral resources under her effective ownership. The minerals were initially owned by the Australian people but rights to exploit them were acquired by her father, Lang Hancock. This is a major story in its own right. An important thing that is not well known is that Hancock did not discover the iron ore – it was discovered by a geologist in the employ of the Western Australian colonial government, Henry Page Woodward. Prima facie, there was no reason to give the ore to Hancock. A counterfactual of some interest: imagine if the ore had stayed in public hands and the profits had been invested in a sovereign wealth fund. What would that fund be worth today?

Offset matters: the bird

With this background, let us turn to the matter of offsets if the Adani mine goes ahead. We commence with a surprise. The very important direct offsets (habitat loss) the company has to put in place do not have to be fully developed before mining commences. However, a finch management plan has been accepted by both the Commonwealth and Queensland governments. Whether those offsets can save the black-throated finch will only be known after the fact.

Let us assume that there is suitable habitat for the bird on land adjacent to the proposed mine and that the birds were successfully translocated there – we will put aside the practical measures required to achieve that. If that was the case, we would have all the proof needed to know that the offsetting worked. But this is not the case for the Adani mine.

If approval for the habitat-destroying project is given before there is proof of success in offsetting the loss of finch habitat, there is a danger of rendering offset policies and laws useless. That approval has been given. Failure of the offset, if it happens, is an unintended outcome of the Adani story. The Adani case is likely to determine the future of biodiversity offsetting in Australia, and have ramifications in other countries given that they look to Australia as a world leader.

Climate change

As noted above, the case of the finch is only one element of the Adani environmental offsets story. The public protests and political campaigns against the Adani mine, ongoing as this is written, are in the main an objection to the addition of greenhouse gases that will result when the coal is burned. As noted above, this will take place in India. Why, then, is there a concern in Australia about this coal mine?

The argument by opponents to the mine is that if it is approved and mining takes place, other very large mines in the Galilee Basin will follow. It sets a precedent. If the others were to go ahead, there would be a very significant increase in carbon dioxide globally. It matters not where the coal is burned.

This potential outcome is considered in terms of damage to the Great Barrier Reef. While this argument is made by coral scientists, it is also playing out in public opinion. It is this fear that has motivated a groundswell of objections to the mine. It is not our task to elaborate on this issue as it has been the subject of extensive reporting and we have nothing new on that score to report – our analysis of the Adani mine case focuses on offsetting, not the Great Barrier Reef.

Offsetting in a dynamic environment?

What offsetting entails has been spelt out in an earlier chapter, but it is useful to elaborate here on a particular aspect. We have noted that offsetting means there is to be no net loss. We can express this in other terms: offsetting means to maintain a steady state, or some would say equilibrium. Each of these terms has its own problem if taken literally. Confining the discussion to biodiversity, we understand the dynamic nature of the natural world; in other words, the baseline is capable of change without any human interference related to the project where offsetting is required. Forest succession is an example. With natural changes occurring, what are we planning to achieve with offsetting?

It pays not to be overly concerned about this in most practical cases of offsetting – we could end up with analysis paralysis. However, the potential of a changing baseline needs to be kept in mind. For example, serious climate change unrelated to the project could have significant baseline-changing impacts and we are then confronted with a different environment from the one which was our baseline for no net loss.

The Outback: no longer out of sight or mind

Our case study is set in the Galilee Basin. This is grazing country, once dominated by merino sheep, now cattle and some sheep. It is brigalow country, part of the much larger Brigalow Belt. We discussed this in Chapter 1, but elaborate here. The dominant plant species – the species that provides its name to this landscape – the brigalow tree (botanists would call it a shrub), is *Acacia harpophylla*, a black-stumped, silver foliaged, hardy wattle. An enormous amount of this was cleared after World War II as brigalow and associated sparse eucalypt country was opened up to become grassland. A controversial Premier of Queensland, Joh Bjelke-Petersen, made his name by using ex-army vehicles dragging a large metal ball through the brigalow scrub to flatten it. The little brigalow that remains has a special protected species status.

The proposed Adani mine is in Banjo Paterson country, think 'Waltzing Matilda'. This unofficial national anthem of Australia was written on Dagworth Station. The nearby (in Outback terms) small township of Winton is where Australia's national airline, QANTAS, was formed. Another not too distant town is Barcaldine, where the Australian Labor Party came into being, formed by striking shearers in 1891. They met at Barcaldine, under the Tree of Knowledge. The more we think about the location of the Adani mine, the unique class history of the area and its natural environmental features, the more we understand that this project is for the history books, whether it goes ahead or not.

Parts of the Galilee Basin are semi-arid and, to make the point that it is near-desert country, just to the north-west of the proposed mine there is a large area known as the Desert Upland Bioregion. Yet there are areas of good Mitchell grass to the west. The water courses in the area flow intermittently. Underground is the Great Artesian Basin, of which Banjo Paterson wrote 'if the lord won't give us water we will get it from the devil deeper down' (in 'Song of the Artesian Water').

Few people live in the Galilee Basin. The pastoral properties tend to be large, the towns scattered widely throughout the region and small in size. The largest township by far is Emerald, with a population of 13 532 in the 2016 census. It is to the south-east of the proposed mine. About equidistant to the south-west is Longreach, with a population in the 2016 census of 2970. Moranbah to the north-east, a town built around mining rather than the pastoral industry, had a population of 8735 in 2016. The nearby town of Clermont had 3031 people in the same year. The mine site is ~160 km north-west of this town. Scattered around the Basin are tiny towns such as Torrens Creek, Muttaburra and Adavale with populations of 50–150 residents.

The approval process commences

Our story commences in late 2010 when the Queensland government (always keep in mind that it is this government that owns the coal on behalf of the people of Queensland) declared that this mine proposal required scrutiny at the highest level; that is, by the relevant Queensland Minister of the Crown and the Co-ordinator General, the latter considered by some to be the most powerful position in the Queensland government bureaucracy. The mine and associated private rail line was declared a significant project in November 2010, a designation which meant that a full-blown environmental impact statement (EIS) was required as the initial step in assessing the Adani proposal.

As the area to be mined included animals and plants deemed endangered under Commonwealth legislation (the *Environmental Protection and Biodiversity Protection Act 1999* [EPBC Act]), the Commonwealth had to agree if the project was to proceed. The fact that the coal is to be exported means that the Constitution brings this government into the picture, as does the fact that a foreign corporation is involved.

The Commonwealth term for a project that involves this level of government is a 'controlled action'. Under a formal agreement between the Commonwealth government and the states and territories, only one EIS is required, assuming it meets the prerequisites of both levels of government. And so it was in 2012 when Adani Mining Pty Ltd submitted its EIS.

Throughout the ensuing years, both the Commonwealth and Queensland governments would be a tag team in assessing and finally approving the project – on conditions. A timeline will help outline the story (see Box 12.1).

Offsetting the depletion of a finite resource: coal

Coal, for all practical purposes, is a finite resource – keep digging it up and burning it and one day it will all be gone. Putting its exploitation in economic terms, deposits will be

Box 12.1: The Timeline

2010

22 October: Adani, supported by the firm GHD Consulting, provided an Initial Advice Statement to the Queensland government.

26 November: The Queensland government determined that the Adani mine and rail line was a 'coordinated project'.

2011

6 January: The Commonwealth government determined that the mine and rail project were, as a package, a 'controlled action' pursuant to the EPBC Act 1999.

25 May: Terms of Reference (ToR) were finalised by the Queensland government (per the Coordinator-General).

2012

5 November: Adani submitted its EIS.

15 November–11 February 2013: Public perusal of the EIS.

2013

26 March: The Queensland government (per the Coordinator General) requested further information from Adani to address matters raised in the EIS.

9 July: Adani applied to the Queensland Environmental Protection Authority for a site-specific Environmental Authority for the mine.

25 November–20 December: Supplementary EIS (SEIS) made available for public comment. Following that, Adani supplied an Additional Information EIS (AEIS).

2014

7 May: The Queensland government (per the Coordinator-General) releases its evaluation of the EIS.

7 May: The Queensland Coordinator-General recommended approval of the mine subject to conditions and recommendations.

24 July: The Commonwealth Minister for the Environment granted approval for the mine and rail line subject to conditions. NB: This approval was withdrawn and a new approval with conditions was granted on 14 October 2014.

28 August: A Draft Environmental Authority was issued to Adani under the Queensland Environmental Protection Authority. Conditions were attached.

1 October: Following public notification of the mining lease application and the application for an associated Draft Environmental Authority, objections to both were lodged. The objections were referred to the Land Court of Queensland on 1 October 2014.

2015

15 December: Land Court of Queensland delivers its findings and recommendations.

2016

3 April: The Adani mine and rail project was approved by the Queensland government, announced in a media release by the Premier and the Minister responsible for mining. NB: this approval was conditional on certain matters being resolved, in particular, a satisfactory offset management plan for the black-throated finch.

2018

9 December: The Commonwealth government signs off on a management plan for the black-throated finch. Conditions are attached.

2019

9 May: The Commonwealth government signs off on a water plan. Conditions are attached.

22 May: The Queensland government signs off on a revised plan for the black-throated finch. Conditions are attached.

12 June: The Queensland government signs off on a revised water plan. Conditions are attached.

available at increasing cost to mine, but when it costs more to extract the coal than it can be sold for it is no longer of use to us. Well before then, alternative sources of energy will be much cheaper, and coal will be left in the ground. Australia reached that stage in 2019. In Australia, both solar and wind farms produce electricity at lesser cost than new coal-fired power stations and, as the intermittency of electricity supply from these sources can be overcome by pumped-storage hydro-electricity, the future would appear secure for these non-polluting, renewable energy resources.

Notwithstanding the case for the renewables, the reality is that we are some years (decades?) away from the end of coal as a source of generating electricity. It takes time to build completely new electricity infrastructure. Although solar and wind farms can be built quickly, there are processes of seeking approval, arranging contracts for the electricity and obtaining finance. All take time. Building large pumped-storage hydro-electricity dams and installing the plant is a longer process. And this still leaves the task of upgrading the major electricity grids, a must as more and more renewable energy enters the grid. It is early days yet.

Let us not bury coal in the sense of writing its obituary without recognising the enormous difference it has made to the human condition. Barbara Freese, a Minnesota lawyer, published an excellent book titled *Coal: A Human History*. She writes: 'Coal was no mere fuel, and no mere article of commerce. It represented humanity's triumph over nature – the foundation of civilization itself' (Freese 2003). She explains how coal provides us with light, power and wealth, and makes for a civilised lifestyle. Darkness, weakness, poverty and barbarism are no more. With one potentially civilisation-destroying consequence omitted, Freese is correct. Today, as we approach the end of the coal age, humanity faces a threat not before experienced in historical terms. This is the climate-changing ability of a build-up of greenhouse gases in the atmosphere. Coal is one of the fossil fuel culprits.

Without delaying the story, we will take a paragraph to make abundantly clear that we are not ignorant of the dramatic climate changes that occurred in prehistoric times, before coal was to blame. We know of these climatic changes through the use of the most sophisticated measurement tools (for example, those capable of drilling into ice cores and providing samples for analysis). Based on the best available science, we can describe with a high degree of certainty the dramatic climatic changes of the past, including the relatively recent thaw that raised sea-levels so high that Tasmania became separated from mainland Australia. These changes were not in any manner related to what humans did.

The most fundamental differences between the past perturbations in climate and the present threat from the build-up of greenhouse gases will be obvious to readers who are familiar with history. For example, when most of northern Europe, including all of Scandinavia, was covered in immense ice fields and when Tasmania was not separated from mainland Australia, the human population was tiny compared to the present. When the ice on land commenced melting, there were no cities with tens of millions of residents perched on oceanic coasts, there was no agriculture spread around the fertile plains of the planet, there was no human-made infrastructure; in fact, our ancestors were of and in nature, basically hunters and gatherers who could easily move in concert with climatic changes. Adaptation was very easy. When a major change in climate threatened, because our ancestors existed in very small numbers they were able to shift to areas where they could re-establish their nomadic lives. It was a world so dramatically different to what we know and experience that it is extremely difficult to reconstruct in our minds. Today, change the global climate to a significant extent and much of what humans have built in the last few hundred years is threatened and, in the worst case, destroyed.

Intergenerational equity and offsetting

We will come to the issue of offsetting the environmental damage caused by burning coal, but the next offsetting issue is what the present generation does – burning coal as if it were a renewable resource and each deposit was replenished after use. In this context, we need to ask if we are morally and, we could add, legally obliged to hand over to future generations a world as productive and pleasant as the one we live in.

We know the coal resource is being depleted. We are reminded daily of the benefits we obtain from burning coal: electricity in the home, lighting and air-conditioning in our places of commerce, energy-saving machinery in our factories. Take coal away and, without a replacement source of power, human society races to a bleak and brutish future. Of course, it is not just coal – the burning of natural gas and oil in its various forms adds layers to the atmospheric blanket which is building the increasingly powerful greenhouse around Earth.

The question is simple: how do we compensate our children and grandchildren if we have burned all the available coal before substitutes are in place, capable of producing 100% of our energy? Before answering that, some might ask, do we have to compensate future people? This is an ethical question, which each person could answer on an individual basis. However, some time ago we answered the question on a collective basis. On our behalf, our governments enacted laws that required adherence to what we now term ecologically sustainable development. These laws answered the question of fairness to the future. The overriding principle was intergenerational equity (alternatively called intertemporal equity). This means nothing more or less than treating each generation equally. This poses a problem if an early generation uses up all the resources a future generation needs to be on par with the previous one.

Before we burn the last tonne of coal which we are going to burn, we need to have in place an alternative power source. This will mean ensuring that the alternative is fostered if it is lagging in development. There are two conflicting views on how this is resolved. First, we feel that there is a cadre of economists (still dominant in the Commonwealth bureaucracy and some state government bureaucracies) who blithely say 'don't worry, the market will ensure that alternatives come on line as needed'. Some may add 'why concern ourselves with future generations as they are going to be better off than we are?' These views are based on the indisputable fact that year-in, year-out industrialised countries have made enormous progress since the Industrial Revolution and there is a belief that notwithstanding the economic damage that climate change is likely to cause, there will be positive economic growth in the future. On this thinking, the damage costs of climate change do no more than slow economic growth. This thinking helps explain the complacency of some economists.

These economists measure progress in Gross Domestic Product (GDP) per capita, notwithstanding its fundamental shortcomings. We need to put to rest this erroneous thinking. One example should suffice. In GDP terms, the greater the damage done by severe cyclones, hailstorms, heatwaves, droughts, coastal erosion and spread of disease (all features of global warming) the better off we are! The huge costs of replacing damaged assets add value to GDP. In a perverse way, measured by GDP, damage from climate change is a benefit.

In coming to a rosy conclusion about the future, we are neglecting the potential serious adverse impacts on our future well-being from climate change. If the impacts are as serious as the official predictions indicate, the next generation and those who follow will be worse off. One question (there are others) is – how do we compensate for the fact that we, selfishly, have used up future generations' major source of energy?

The realistic economic solution: Hartwick and sovereign wealth funds

Here we introduce the Hartwick Rule. In plain English, this rule states that if a nation's well-being is to be sustained through time at its existing level it has to replace natural capital, such as a coal body and associated mines, with another capital asset capable of producing the same level of well-being. If the asset to be replaced plays a unique role, such as coal does in the generation of electricity, the substitute asset must do the same thing. This is the concept of like-for-like, in offsetting jargon. This adds a special dimension to the role of a sovereign wealth fund.

In practical terms it means that the 'profits' (using this term to cover resource rents, royalties plus conventional profit) earned from the exploitation of coal have to be invested in other economic pursuits capable of earning the same level of income as coal mining did, with the proviso that a proportion of the investment needs to be directed to the replacement of coal-fired power stations. The obvious candidates are hydro-electricity schemes, including pumped-storage hydro, wind turbines, geothermal power systems and large-scale solar farms. Other energy sources could eventuate, in particular nuclear fusion and hydrogen-based energy, but they remain on the drawing board rather than knocking on the door.

As solar and wind farms are standing on their own feet in cost terms and hydro-electricity is a mature industry with pumped-storage power compensating for the intermittency of wind and solar, we can expect that the renewable energy sector will displace the fossil-fuel-derived stationary energy sector without specifically targeted investment. Consequently, the profits obtained from mining and burning fossil fuels can be invested in a wide portfolio. Of this, we have real-world examples. One sticks out. Paul Cleary, a senior writer for *The Australian* newspaper, tells the story.

Cleary commences by comparing Australia to Norway, as these countries are number one and two on the UN Human Development Index. Norway pips Australia. Nature has been very kind to both countries – Norway has oil to export, and Australia has coal to export.

Cleary (2011, p. 64) writes:

… a combination of resource wealth and first-world institutions has proved to be the sweet spot, evaluating quality of life in the two countries to the maximum available. But Norway differs from Australia in one very important way – it is certain to remain in its enviable position long after the resources have been depleted. Australia has no such certainty. Norway … has … had great leadership, including that of former Prime Minister Gro Harlem Brundtland … Brundtland articulated the concepts of sustainable development and inter-generational equity – the idea that the current generation should not leave future generations worse off … Taking on board the principles of inter-generational equity, [the Norwegians] set up a petroleum fund that is as breathtaking as it is simple in design and purpose. In essence, Norway's fund transforms a non-renewable resource into a financial asset that can last forever. It does this through the principle that no more than … four per cent of the fund should be spent in any one year.

There are other important features of Norway's sovereign wealth fund. A vital one is that the oil profits are invested overseas, not in Norway. As a consequence, there is no inflationary impact in Norway. Another feature is that investment is not permitted in ethically unsound industries.

Norway's sovereign wealth fund is boosted by government majority ownership of its oil, and an additional tax on oil production. When Cleary published his book in 2011, the extra tax was set at 50% on the normal corporate tax rate. This is justified 'for the privilege of profiting from the wealth of Norwegians' (p. 66). While we can read in newspapers and hear on talkback radio and current affairs television programs that there are calls in Australia for the introduction of a sovereign wealth fund and an ownership position with regard to the underground resources that Australians already own, the government appears disinclined to take these economically and ethically justified actions. Cleary (p. 71) found that economists in the Commonwealth Treasury had 'a curious bias against sovereign wealth funds'.

The power of the mining lobby and the media that support it in Australia appear simply too strong to introduce a genuine resource rent tax. That there is very significant foreign ownership of mining and energy businesses is conveniently overlooked. Where do the profits go? From where is the mine machinery imported? As Cleary (p. 87) states, 'Trying to establish just how much money the [mining] industry makes is no longer as straightforward as it once was'. He makes the point (p. 83) that 'taxes and royalties haven't kept up with the huge profits now being earned'.

The obvious solution would be for Australian governments to retain ownership of coal and other underground resources and invite mining companies to tender for the task of mining the resources, for a realistic return. This is not contemplated, notwithstanding evidence that the approach works well in other countries. The case made by Cleary is likely to be lambasted as 'socialism' by mining interests and the conservative politicians who support the miners. The irony is that the Norwegian sovereign wealth fund was introduced by a conservative government.

In summary, Australia is not concerned with offsetting the depletion of non-renewable resources – and this applies to the proposed Adani mine. It is not at all clear how the nation will meet the inter-generational equity principle to which it has signed up. Not just in the Adani case, but generally, we are failing to offset the declining wealth of non-renewable resources.

Offsetting the greenhouse gases emitted when the coal is burned

The burning of the coal mined by the Adani company will produce a considerable amount of carbon dioxide. An estimate that gets considerable air-play (based on calculations made by The Australia Institute – Cleary 2011) suggests that the Adani coal, when burned, will create annual emissions of a similar order of magnitude as countries such as Malaysia and Austria, more than Bangladesh and Sri Lanka, and only marginally less than Vietnam. We know that the Adani coal will fire power stations in India. These power stations, more appropriately the ultimate consumer, electricity-using households and businesses in India, are responsible for the greenhouse gas emissions. This fact has not stopped many Australians from arguing that the mine not be allowed. Or if allowed, it should be on the condition of sequestering the carbon dioxide produced. It would be a mammoth effort to, say, plant forests to compensate for the extra emissions but it could be done. Who would do it?

Those who argue that Australians have that responsibility base their case on an interpretation of the EPBC Act. For example, is the definition of the environment in this Act broad enough to apply to the global environment? Or – basically the same question – is the planetary ecosystem consistent with the definition in section 528 of the Act? Social, economic and cultural matters are defined as part, or aspects, of the environment. In as much as we live in a global economy – of this we are reminded by politicians and media outlets – economic matters are global in nature, and mining and burning coal is an economic activity.

Notwithstanding conflicting interpretations of the Australian law, international agreements require emissions to be counted in the country where they occur. This is the polluter-pays principle in operation. In climate change jargon, these are called Scope 3 emissions.

Those making the case against the Adani mine argue that its contribution to climate change will be disastrous for the health of the Great Barrier Reef. Professor Ove Hoegh-Guldberg was asked to address this matter in a legal challenge to the Adani mine (Land Court of Queensland 2015). In summary, he found that:

> As we are already above the thermal threshold for damage to reef building corals and hence coral reefs, any further addition of CO_2 into the atmosphere will directly damage the Great Barrier Reef ... The true cost of the emitted carbon from the Carmichael Mine to the Great Barrier Reef and other ecosystems, businesses and human health must be calculated and attached to any decision on whether or not to proceed with the mine. To ignore the impact of the mine, knowing that the emissions from the extracted coal are not going to be sequestered, ignores the much greater costs of the mine to people and businesses worldwide.

Hoegh-Gulberg is correct in stating that there is no proposal to offset the extra greenhouse gases entering the atmosphere as a result of burning the Adani coal.

We should note that sequestering is used to describe two radically different processes. The usual process, the one promoted to airline passengers, is to grow trees, which take in and store carbon dioxide. Airline passengers can make a voluntary contribution of a few dollars to an organisation which undertakes to plant trees. Trees and all matter of living things eventually become coal. However, trees planted today decay and in the process release CO_2. Hundreds of millions of years would have to pass before the decayed trees became coal.

The other method of sequestration will, if it becomes economically viable, entail capturing carbon dioxide as it comes out of the power station and transporting it to a disposal site where it will be stored permanently. Various homes for the CO_2 have been suggested, including abandoned coal mines and the deep ocean. The evidence to date suggests that this form of carbon sequestration is technically feasible but immensely costly and probably decades away. By then the coal age is likely to have ended.

In summary, the question of offsetting the increased greenhouse gas emissions from the Adani mine is a major matter, but if anything is to be done in this regard it needs to take place in India and be paid for by Adani. The view that the polluter is the consumer not the producer is too well established to be denied. In terms of jurisprudence, the opposite view as to the culprit is a matter that needs to be sorted out. All we can do is note the diametrically opposed positions.

Offsetting direct impact: endangered species

The Adani mine and rail line will have impacts on 48 defined environmental values. Most importantly, there are threatened and endangered species in the area to be mined. The mine site is visited by migratory species. One ecosystem, one plant and one animal species – the local brigalow ecosystem, the waxy cabbage palm and the black-throated finch – are listed under the EPBC Act as endangered. Offset plans are required for each under the conditions of approval granted to the mining company by both the Commonwealth and Queensland governments. In addition, the considerable amount of bore water to be taken by the mine has to be made good (in our language, offset) so that local pastoralists are no worse off in water availability.

Here, we will summarise the situation as it stands today. For an extensive analysis of the Adani mine proposal and the offsetting required, *Adani versus the Black-throated Finch* (Hundloe 2018) is the only source currently available. After nearly a decade of Queensland and Commonwealth government assessment of the project, numerous consultants' report and court cases, two offset matters have not been resolved on the ground; that is, in practice. In making this statement, we acknowledge that management plans have been approved by both levels of government for both the finch and the use of waste. Put simply, a management plan is no more than a plan. What happens in practice – on the ground – is the proof of whether offsets are achieving their stated aim. It is too late if the plan does not deliver the outcome expected of it. It is important to note that it is possible to have confidence in some offset plans and not in others.

- Has this approach been applied before?
- Is it possible to point to examples of success; for example, where like-for-like environments have been located and secured?
- Where habitats have been restored or re-created, how has the delay involved been addressed?

The answers to such questions indicate whether an offset plan is likely to deliver what it promises. If the answer is 'no' to any of these questions, we are in the realms of uncertainty. Uncertainty arises when there is scientific uncertainty – we might not know enough about the plant or animal or we might not know enough about the ecosystem, such as whether there is an underground aquifer. The fact that it took a very long time for governments to agree to the offset plans and, when they did, this was a result of political events, does not suggest a high degree of confidence in success.

Uncertainty and falling back on monetary compensation

Monetary compensation is not involved in the Adani case but it is very popular in Queensland, including for another mining case. The popularity of monetary compensation means it deserves comment. To be adamant, monetary compensation is not justified in certain circumstances. This is obviously the case when we are seeking to offset the loss of habitat for an endangered species or damage or destruction of an ecosystem. Take a hypothetical worst-case scenario: how much money would compensate for the extinction of a species? Whatever the sum was, how could we arrive at a like-for-like outcome or a no-loss result – clone the species?

Monetary compensation is allowed by both the Queensland and Commonwealth governments when it is impossible to meet on-the-ground, like-for-like offsets. However, this approach has (or should have) very limited scope. The first problem, and it is a substantial problem, is that the damage done by a project has to be measurable in monetary terms. Damage to some valuable assets, for example good-quality soils or grazing country, can be readily calculated. It is the monetary value of the lost production for the length of time the land is unviable for farming, assuming the land will be restored in due course.

Now consider the black-throated finch, the bird listed as endangered and subject to dramatic loss of habitat over past decades. Until recently, the finch and cattle have been sharing the Adani land and suitable land on neighbouring properties. Given the large flock of finches living off this land – finding the seeds of their choice, water in close proximity and nesting sites – their co-existence with the cattle herd is a fact. Cattle do restrict the finch population due to grazing and trampling, and we surmise that without the cattle the carrying capacity of the land for the finches would be considerably greater than it is at present. The present is what matters.

Now introduce mining. As more and more of the Adani land is cleared for mining, the co-existence of cattle and finches breaks down. Clearing results in less and less land for cattle and finches on the Adani land, a former grazing property purchased by the Adani Group. The cattle, in all probability, would be sold. This would 'create' finch habitat on the land previously grazed by cattle once it had the chance to restore itself. The thesis was that finches were not using all the Adani land available to them because grazing and trampling were depleting the finches' grass-seed food. Restoration of once-grazed land would solve this issue.

With this scenario, the best-case situation would be immediate restoration of the land once the cattle were removed. This could mean nothing more than seeding the grass which is the preferred species for the finch. Within a relatively short period there would be additional habitat for the existing finch population. With sensible planning, this transformation of the land could commence before any of the presently available finch habitat was lost to mining. The finches could occupy this land before they lost the habitat they had before mining. The crucial matter is to eliminate any time lag between loss of presently used finch habitat and the restoration of replacement habitat.

Another scenario is that there are properties adjoining the Adani one which have areas of suitable finch habitat. This is true. We would find an extant finch population using that land, and expect that finch population to be in line with the carrying capacity of that habitat. Assume it is possible to translocate finches from the Adani mine site to these neighbouring properties as they start to lose habitat due to the commencement of mining. Would this work? Would there not be intra-species rivalry, competition for food and shelter? We would not expect this translocation to be a viable option as we have done nothing to increase the carrying capacity of the land on the neighbouring properties.

The third scenario is that there are no, or only a sub-optimal number, of finches on the neighbouring properties, because grazing has diminished the extent of their favourite grasses. The possible solution would be to reduce the cattle herd and seed the area with the finches' preferred grasses. The result would be restored finch habitat, and the only issue would be to entice the finches to relocate from the Adani mine site.

If none of these scenarios are available, for whatever reason (that is, the dislodged finches are not able to be accommodated on the Adani property or neighbouring ones), what happens? No viable solutions are evident.

Consider a counterfactual. The existence of coal deposits in the Galilee Basin has been known for a considerable time. Governments have also known of the endangered fauna and flora in the area. A rational planning approach would have started the search for offset solutions at the time mine exploration leases were sought and granted. A form of advanced offsets could have been in place. It is too late now and a messy situation exists, with no guarantee there will be a winner.

Conclusion

The black-throated finch has lost 80% of its original habitat due to land clearing. It is no longer found in New South Wales. There are only two significant areas where it is found, in the Townsville–Charters Towers region and on the Adani mine site. We could assume that the present population of black-throated finch living on and around the mine site is optimum given the amount of available preferred grass seeds, water and nesting sites. The mining operations will destroy some of the resources required by these birds. All other things being equal, we would predict a decline in the birds' population. Put in other terms, if the birds presently reliant on the area to be mined were pushed into an area not subject to mining but already providing for its own population of black-throated finches, there would be competition for food and shelter. The end result would be a downward adjustment of the total population of black-throated finches. This is failure.

It is possible that some land adjacent to the mine site could be rehabilitated to allow for translocation of the finches. This could require the pastoralists to reduce their herds, eliminate feral animals and reintroduce the birds' favourite grasses. We know from documents available to the public that the Adani ecological consultants have identified properties adjoining the Adani site that have promise for this purpose. We also know, from the same documents, that the possibility of retaining the bird population on the Adani property is being investigated.

However, if we consider the evidence presented in the Land Court of Queensland (2015), witnesses for Adani and for those who oppose the mine admitted that there was a high level of uncertainty as to why the black-throated finches prefer the habitat on the Adani mine site. The same degree of uncertainty bedevils two other offset requirements, the brigalow plant community and the waxy cabbage palm. With regard to the latter, the judge hearing the case, President Carmel MacDonald, said:

> It is clear that Waxy Cabbage Palm is a vulnerable species and that the ... population is the most significant in the world. Beyond that, little appears certain. There is uncer-

*tainty as to optimum conditions that are necessary for the continued likely survival of
… the population. It is also uncertain what impact the mining operation will have.*

We are not in a position to conclude the Adani story. We will not know if the management plans are successful – until we have evidence of happy, translocated birds. We remain sceptical of this possibility.

In a similar case a few years ago, opponents to a project in which offsets were the suggested solution gained access to the Federal Court. The case was *Northern Inland Council for the Environment v Minister for Environment* in 2013. A critical issue was the approval to clear land for mining before the offset details were agreed. The judge commented, 'It is correct … that offset conditions need not be satisfied before commencing approved clearing'. This would suggest a failure to draft appropriate legislation.

If the Adani mine site is cleared and the offsets suggested for the finch subsequently fail, it will be too late to fix the problem. If biodiversity offsets are to have a future, they need to be limited to situations where they are capable of success – and this needs to be ascertained before the start of bulldozer engines. The existing legal position – that a project can start before offset conditions are known and shown to be working – must be abandoned and replaced by a sensible ordering of the approval stages.

THE MAJOR TASK FOR THE FUTURE

An urgent matter: offsetting coal mining jobs

Tor Hundloe and Ella Dewilde

The case studies have illustrated a range of environmental offsetting initiatives. Some have been successful given their objectives, others have been partly successful, and there are serious doubts of success in at least one case. In this chapter, our focus is the future. We address one of the most perplexing issues facing Australia in the early part of the 21st century – the future of work in mining coal. Coal-fired electricity is one of the major emitters of carbon dioxide. As a global matter, there is a direct link to the risks associated with climate charge.

The burning of coal is only one of a range of influences on climate change. Others include the extensive forest fires that are regularly lit in the Amazon and South-East Asia, particular in Kalimantan, Indonesia. There is the burning of petrol and diesel in moving people and goods by car, truck, rail and sea. There are numerous other sources of greenhouse gases. However, climate change *per se* is not our focus. Our focus is rather on one of the consequences of addressing climate change, the cessation of burning coal.

If coal-mining is to be phased out, whether by government policies or a decrease in world demand for coal, jobs will be lost in the industry and the day will come when there will be no jobs left at all – the industry will not exist. This would obviously be of great concern to coal miners and their families, as well as to business and community leaders in mining towns and regional cities where economic links to the industry are strong. As a digression, a recent proposal by the Commonwealth government to bring migrant workers into the nation's regional economies at a time when people in these areas worry about losing employment is adding to their concern.

Recently, we have witnessed a higher level of debate on the matter of coal-mining jobs than on any other environmental issue. For some, the loss, or potential loss, of coal-mining jobs is a surrogate for the debate about climate change. By focusing on jobs, the debate can be more readily framed as an immediate, local issue rather than arguing that human-induced climate change does not matter because in a grand sweep of undetermined time the planet's climate is likely to change naturally, as it has done in the past. This is not the place to enter into a discussion of assertions that the doubling of greenhouse gas emissions since the start of the Industrial Revolution does not matter. It does! We shall leave it at that.

We deal with a less emotionally charged matter. It is how to phase out coal for electricity generation as more and more electricity is provided by renewable electricity sources.

Renewable energy in its various forms is scheduled to dominate the world's energy industry by 2050, if not before. This is occurring because it has become less expensive to provide electricity by renewables, as very significant economies of scale are being realised. Building a new power station in Australia is no longer an economic proposition.

Governments commenced by treating the renewable energy industries as infant industries that deserve subsidies, this is no longer necessary as the industries have matured and can stand on their own feet. However, this does not suggest that subsidies should not be given to the renewable energy sector. There are two conventional economic reasons to continue to grant the subsidies.

The first is that subsidies to renewable energy industries are a partial substitute for the non-existence of an externalities tax (a charge for the eventual damage carbon dioxide build-up in the atmosphere will do to our economy). It is, understandably, very difficult to estimate the level of a carbon tax that would result in reducing carbon dioxide levels to the extent that damage did not occur. While we are aware of the general levels of damage associated with different degrees of global warming, to be reasonably confident of the actual cost of, say, more severe droughts is another matter. We are forced to do the best we can, given uncertainty and unclear risks. It is very much a second-best solution when we decide on the magnitude of subsidies to carbon-free energy sources. From an economic perspective, the best solution would be a carbon tax that dramatically reduced greenhouse gas emissions.

The second rationale for the continuation of renewable energy subsidies is another second-best case: the coal-mining industry is granted subsidies of various kinds, which naturally put it into a favourable position in relation to its competitors. Economic efficiency can only be achieved if all industries are treated equally. The economists' second-best solution is subsidise one industry and subsidise its competitor industries.

These points noted, a case can be made for treating the renewable energy sector more favourably than the coal industry. This centres on the economic principle of inter-temporal efficiency (otherwise known as inter-generational efficiency). The case is as follows. For all practical purposes coal is a non-renewable resource and the economic benefits of using coal today are not going to result tomorrow. It is a once-only situation. On the other hand, the use today of renewable resources such as sun, wind and water do not diminish those resources and the economic benefits of their use tomorrow.

The only solution to address inter-generational equity in the case of an exhaustible resource, such as coal, is to take whatever surplus is earned, over and above the costs of extraction, and invest it in a sovereign wealth fund, as Norway does with its oil profits. We should note that this is easy to do in Norway because the government is the majority owner of the resource and hence faces no opposition to the concept. We do not have a sovereign wealth fund in Australia. Furthermore, it is not at all clear that we, the owners of the coal, are getting a fair price when it is awarded to a mining company. If we were getting a fair price, mining magnates would not be billionaires. Their costs of mining plus a fair profit based on the opportunity cost of capital invested would be the totality of their earnings, the rest should be in a sovereign wealth fund. Such a fund should be capable of earning enough income in perpetuity to offset the income presently earned by mining. This is how an economist would deal with inter-generational equity.

In addition to the cost advantage of renewables, the other force driving the end of coal mining is the imperative to deal with climate change. While the Australian national government, at the time of writing, has no policy on ending the use of coal to generate electricity, other countries do. For example, Germany has set the end date at 2038 and is devising plans for the displaced workforce. The UK leads the world with its roll-out of off-shore wind turbines, and plans to have no carbon dioxide emission by 2050. By 2030, India plans to have 60% of its power generated by sources other than fossil fuels. There are numerous other examples. Regardless of what we do or don't do in Australia, the energy sources of the future will be greenhouse gas-neutral, renewable sources.

It is not just the use of coal to generate electricity that has a limited future. Coal has another use, in the production of steel. This use is also about to be challenged. Serious efforts are being made by some of the world's largest corporations to replace coal with hydrogen to remove oxygen from iron ore to produce pig iron, with water rather than carbon dioxide being the residue.

When postulating a future in which less and less coal is mined for electricity generation, we have to consider more than work in mines. The labour force in power stations must also be taken into account. The existing ones will gradually close, several of them simply because they are old. Notwithstanding the importance of transitioning the power-station workforces, we do not discuss this matter here. However, we must remain mindful of the need to deal with the transition of workers in this segment of the industry.

Offsetting jobs for coal miners becomes of high public policy relevance. With regard to proposals for new coal mines, this matter cannot be put off to some future time. It has already led to major controversy in relation to the Carmichael (Adani) coal mine. Many other potential coal mines are on the drawing board; if any become an actual proposal, it would be controversial and a source of societal conflict. These factors lead us to an investigation of alternative employment for coal-mine workers and of alternative industries capable of offsetting the loss of workers' income, much of which supports businesses and other jobs in local and regional economies. The State of Queensland, as the most decentralised in Australia, tends to be the focus of concern about regional employment.

The logical place to commence our investigation is employment opportunities in the rapidly expanding renewable energy sector, comprising solar, wind, hydro and waste-generated electricity, plus battery storage. This limits our investigation to comparison of employment in the coal-for-electricity segment of coal mining with employment in the renewable electricity sector. Our analysis is confined to the situation in Australia. We are not in a position to analyse the job situation in countries which purchase Australia's coal for electricity generation and smelting iron ore. Furthermore, we are not able to throw light on the employment generation in countries manufacturing solar photovoltaic panels and wind turbines. Only potential job losses and job gains in Australia are our subject matter.

A few coal-mining facts

Before proceeding to the heart of our investigations, we should note a few facts about Australian coal. On a global scale Australia is a significant producer of coal, and in terms of exports it provides about one-third of world imports. Approximately three-quarters of Australian coal is exported. Most of the balance is used for domestic electricity production. The coal-mining states are Queensland, New South Wales (NSW) and Victoria. Both black (bituminous) and brown (lignite) coals are mined. The more valuable black coal is found in Queensland and NSW.

There is a difference between-coal-for-electricity (thermal coal) and coking (metallurgical) coal. Australia produces much of both. Coking coal is used primarily in the smelting process, where heat is applied to ore to extract the base metal. By this process we get iron. Thermal coal is crushed, fine-grained coal that is burned to turn water into steam, and the resultant high-pressure steam is used to spin a turbine connected to an electricity-producing generator.

The point of commencement for our investigation is to gather facts on employment in coal mining, both pre-mining and operational work. The work in preparing a mine and putting in place the necessary infrastructure entails different types of work from that in operating a mine. Our primary interest will be in the operational stage, as this is of the longest duration by far. Coal mines can have lives of up to 60 years at the present rate of extraction, although the indications are that a 30-year lifespan is probably the maximum unless carbon capture and storage become economically feasible and coal continues to be used to produce electricity. There is also a post-mining phase which comprises land restoration and related decommissioning work. Except for periodic checks on the progress of restoration, this phase is not labour-intensive once the land has been re-contoured and plants established.

It might be assumed that it would be an easy task to ascertain the number of people employed in coal-mining. That assumption would be wrong. The Australian Bureau of Statistics is tasked with the job of counting jobs. It uses two different methods – a reasonable approach – however, they result in different numbers. We will go with the estimate of 38 100 coal-mining jobs in 2017–18, while noting there is an estimate of 52 600 jobs in 2019 when using the less-preferred method. There is no breakdown in employment between thermal and metallurgical coal; however, for some time the ratio has been ~60:40, with a most conservative estimate of 53:47. On this basis, we expect the number of workers presently engaged in thermal coal mining to be 21 000–29 000.

Coal mining is concentrated in certain regions of the nation. The major areas are, from north to south: the Galilee Basin west of Townsville, the Bowen Basin west of Mackay and Bowen, the Surat Basin west of Maryborough, the Clarence-Moreton Basin west of Brisbane, the Sydney Basin west of Newcastle and the Gippsland Basin in the Latrobe Valley. Scattered throughout these regions are mining towns. Some were specially built to service the local mines while others are regional centres in their own right, such as Emerald in central Queensland which services a large agricultural sector as well as mining. As modern mining is a mix of on-site workers and fly-in-fly-out workers, major cities in the general area can be home for workers. In Queensland, Townsville, Mackay and Rockhampton are important in this context; however, workers live as far away as Brisbane and the Gold Coast. In Queensland, there is a propensity for wealthier workers to purchase holiday houses in the coastal cities – fishing trips to the Great Barrier Reef in the non-working weeks are a significant attraction.

The variety of jobs in the coal-mining industry is extensive, from highly skilled engineers, managers and accountants to a wide range of lower-level technical skills – all the technical jobs require TAFE certification. The top three occupations are categorised as drillers/miners/shot-firers, metal fitters/machinists and technicians. In addition, there are dump-truck drivers, plant operators, excavator operators, electricians, mechanics, dogmen, riggers, fitters, mining surveyors, process engineers and analytical chemists. Another group includes occupational health and safety officers, environmental officers, supervising engineers and, finally, managerial and clerical staff.

While the list of occupations is extensive, the number of workers involved in mining is not. Mining of most sorts, including coal mining, is increasingly capital-intensive. Automation and robotics are eliminating jobs. For example, driverless trucks have found their way into the mining industry. The world of work is changing! This is recognised at the highest level: on 10 January 2017, US President Barack Obama made reference in his farewell speech to the 'relentless pace of automation'.

An indication of the capital intensity of coal mining can be illustrated with reference to the proposed Carmichael coal mine in Central Queensland. In presenting evidence to a court in Queensland, an economist representing the mining company said that there would be just under 1500 full-time permanent jobs involved in the operational stage of the mine, if the mine was worked for 30 years or 60 years. In contrast, the capital expenditure would be A$16 billion. There would be less than one job for every A$10 million invested!

Is there a ready-made solution?

The Climate Council has had research undertaken which suggests that if Australia were to install renewable energy sources to deliver 50% of its requirements by 2030 there would be 28 000 new jobs created. This was cause for Jennifer Rayner, in her 2018 book *Blue Collar Frayed: Working Men in Tomorrow's Economy* to claim 'renewable energy is the no-brainer blue-collar growth industry.' We certainly agree with her, and make the very important point that much of the growth in these jobs will be spread around regional Australia. Figure 13.1 shows existing and approved renewable energy projects (mainly solar and wind farms) throughout Australia. There is significant overlap with the coal basins of the nation, a factor of major importance when regional employment is a contentious issue.

There is a significant difference in terms of the demand for labour between coal mining and the renewable energy industry. In the latter, the major demand for labour is in the construction phase, while in mining there is a relatively larger workforce in the operational phase. Solar and wind farms and hydro-electricity plants do not require a large workforce once they become operational. However, the very significant expansion of the renewable energy sector is expected to be spread over decades and, consequently, a substantial workforce will be needed for a lengthy period. This compensates for the different operational workforces in the competing industries.

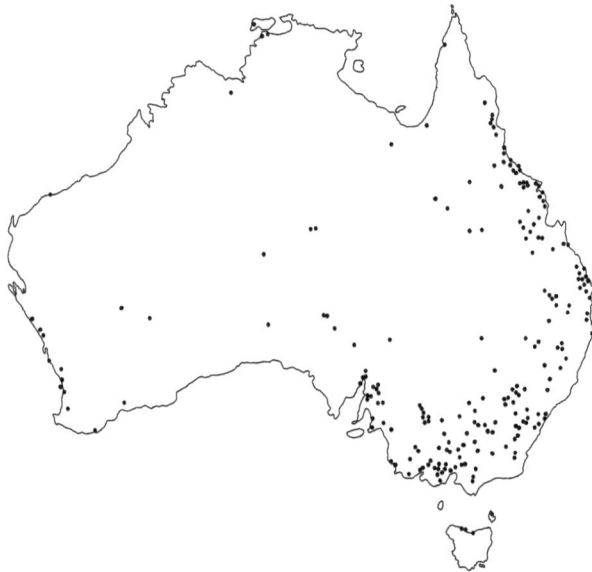

Fig. 13.1. Locations of renewable energy projects that are operating or under construction in Australia. Source: Map by Tor Hundloe based on data from Australian Renewable Energy Agency; Clean Energy Council; National Wind Farm Commissioner; state and territory governments and individual renewable energy firms.

In addition to the construction of solar and wind farms plus hydro-electricity schemes (such as the well-known Snowy 2.0 pumped-storage proposal), there will need to be a very substantial upgrade of existing electricity grids (networks) and inter-connectors to cater for the new technologies. Grid upgrading is particularly important where the new electricity sources are distant from major cities. At present, the wires stretching to inland small towns and rural communities are not of the necessary standard. Again, much work will be in regional Australia. In rebuilding the major grids, there are the intermittencies of both solar and wind power to be dealt with. With widespread pumped-storage hydro-power facilities (yet to be built) the intermittency of production of electricity is overcome if the necessary co-ordinating systems are put in place. The rather rapidly developing battery technology will soon play a major co-ordinating role, and the battery industry will involve an increase in demand for workers.

Overall, there is much to be done in the next few decades in taking the renewable energy revolution to full-scale maturity. There is scant available evidence to suggest that the relevant authorities (governments and training institutions) have considered this in any detail, or particularly investigated the workforce required to facilitate a total renewable energy future. To estimate the total workforce required on an annual basis, and over how many decades, are tasks that cannot be answered without knowing the rate of installation of the various renewable technologies. It is not just a matter of estimating the total number of workers who will be required, but rather the numbers with the requisite skills. We know that electricians are going to be in demand but other skilled workers will also be needed, and the skills are different for each industry.

In 2016, the Climate Council commissioned a study to estimate the additional jobs that could be created in the renewable sector compared to a business-as-usual case. Unfortunately, the economic modelling is not transparent; however, several of the findings and observations appear to be realistic. The study authors suggested that more than half of the new jobs will be in the installation rooftop solar panels. As this will mainly occur in cities and towns, suitably qualified workers will be drawn locally and hence regional employment will not benefit to a significant extent.

The remainder of the new work will be in the large (utility-scale) construction of solar and wind farms, and pumped-storage hydro-electricity generation. This work will occur outside the major urban areas – some of it in the outback – providing a boost for regional communities. For work on a continuing basis in the construction of the new infrastructure, it is likely that teams of workers will need to move from site to site – not that different from the teams of shearers who, for approaching 200 years, have moved around the nation according to the availability of work. An important insight from the research is something that we have noted – there will be both matches and mismatches in skills if the workers entering the renewable sector are those displaced with the closure of coal mines. This will necessitate a planned approach to the shift in jobs, and retraining where necessary. This is a fundamental point we have made.

Planning the transition

Having established the number of workers who will lose their jobs in the coal-mining industry, the next task is to develop a profile of their professional or technical skills, formal qualifications, age, personal circumstances (married, family numbers, working partner, etc.), their home base and their aspirations when mining work ceases. On analysis of these data, the displaced workers' likelihood of finding alternative – offsetting – work can be ascertained. As discussed above, the possibility that many will find work in the renewable energy industry is not wishful thinking. Of course, if there are not large-scale renewable energy projects in the area where mines are to be closed, there will need to be other industries that are expanding if the transition from mine work to other employment is to be smooth. This requires a systematic analysis of the local economies, a task that has to be undertaken in concert with the study of coal-miners' skills and aspirations.

We need to point out that not every outback regional economy is going to stay the same indefinitely. Extremely important efficiencies have occurred in much of Australia's agricultural sector. While very valuable to the agricultural sector, these have had a very significant negative impact on employment and a flow-on effect on businesses in regional economies. Those who are concerned with the withering of country towns need to recognise that, for outback towns to again resemble the much more vibrant past, they will have to convince the agricultural sector to return to a much more labour-intensive means of production. Good luck with that!

Beyond the replacement of labour by machines, there is the most obvious fact that Australia's sheep numbers have been halved, replaced by beef cattle. Small towns that benefited by providing groceries and alcoholic drinks to visiting shearers, and weekend accommodation between sheds, get far less custom in the present era. Grazing cattle is far less labour-intensive than running sheep for wool production.

In the transition, the ideal situation would be, in the jargon of biodiversity offsetting, no net loss of like-for-like jobs. For example, a mechanic working in the coal-mining industry needs to get a job as a mechanic in industry X, at the same annual wage as earned in mining and with comparable conditions. If these criteria are not met, offsetting the loss of mining jobs is likely to require retraining workers or paying for their relocation to a place with a favourable labour market. It should be obvious that offsetting jobs in coal-mining is no trivial matter, and has to be planned well in advance of the closure of a mine. In Australia, looking forward 20 or more years, there won't be one coal mine to close but a number due to their age.

We note that there is a hesitancy to undertake workforce planning in Australia, which is a result of the *laissez faire* economic philosophy which has dominated since the early 1980s. This will need to change. Furthermore, there is a propensity for school-leavers to seek entry to a university rather than a trade course, and an oversupply in some popular professional degrees (law is a prime example of an over-sold degree). While learning is an enjoyable and important part of life, learning profession-specific material which is not going to be used is a waste of the student's time and society's resources – far better to study science, history or philosophy or a profession or technical field where there is high demand. The increasing significance of the renewable energy industry will open up opportunities for workers with the technical skills provided by the TAFE sector of higher education.

Job offsetting has been successful in the past

The worker transition procedure outlined above has been applied in Australia in the past, but on an *ad hoc* basis. For it to happen there had to be a stimulus, such as a groundswell of public opinion to encourage the appropriate decision-makers to act. In one well-known case this type of planning for a transition in employment was adopted and was successful. However, a degree of luck was involved as the regional economy was expanding at the time the transition occurred. Obviously, this cannot be counted on.

The case involved the forced cessation of timber-getting in the far north Queensland rainforests. This was necessary if the rainforests were to be listed by UNESCO as a World

Heritage area. The listing took place in 1989. At the start of the campaign to have the rainforests declared a World Heritage area, the matter of job transition was recognised by the environment movement. The Australian Conservation Foundation commissioned a study of alternative job opportunities for timber-getters. This proactive and eminently sensible action took place seven years before the listing.

As the listing drew closer, it was not left to the not-for-profit sector to carry the day on offsetting jobs and support for the local economy. Due to ethical considerations (and conflict between the pro-listing national government and the anti-listing Queensland government), as a prerequisite for acceptance of the proposal to have the area listed UNESCO required an independent analysis of how the timber-workers and the timber industry would fare. If the situation required it, offsetting was obligatory and UNESCO was to be informed of how it was to be achieved.

Briefly, the economic analysis that underpinned an offsetting package (undertaken by the first-named author of this chapter and his colleagues) provided the necessary data on which the national government could develop its policy. The national parliament received the offsetting studies and debated them. The offset package included various sums of money for retraining and/or relocating workers, grants to local government authorities bordering the proposed World Heritage area (so that various minor infrastructure projects could be undertaken and, hence, generate jobs) and capital expenditure to retool a timber mill so it could work plantation timber rather than the rainforest timber it milled previously.

How ready is the renewable electricity sector to offset coal-mining jobs?

As noted previously, a critical feature of the developing renewable energy industry is the wide distribution of projects, a vast number in the regional economies of Australia.

We now turn to the available data on the expenditure and employment in the renewable energy sector.

As of 2018, the investment in renewable energy production in Australia was in the order of A$20 billion. Large-scale solar plants dominate the current investment, at A$15.3 billion. Approximately 13 500 jobs are directly associated with that level of investment, in construction, operation and maintenance. At the time of writing (mid-2019) the investment had grown to A$27 billion.

Jobs created

A 2018 SKM report (Stocks and Blakers 2018) found that each 100 MW of energy generated by renewables creates 96 jobs during the construction phase. In other words, approximately one job per 1 MW of electricity. However, during the operation and maintenance phases, minimal jobs are created. As noted above, this is not necessarily a problem for the foreseeable future – certainly not until the Australian economy has been transformed, and by then replacement of long-established rooftop solar panels and solar and wind farms will be required and demand for skilled labour will continue. The graphs from the Australian Bureau of Statistics demonstrate state and nation-wide trends in employment across the renewable energy sector. Figure 13.2 illustrates direct employment of renewable energy, from 2009–10 to 2017–18.

Figure 13.3 shows employment in the rooftop solar sector of the renewable industry. Various factors underpin the shape of the graph. Alterations in feed-in tariffs are one

FTE employment (no.)

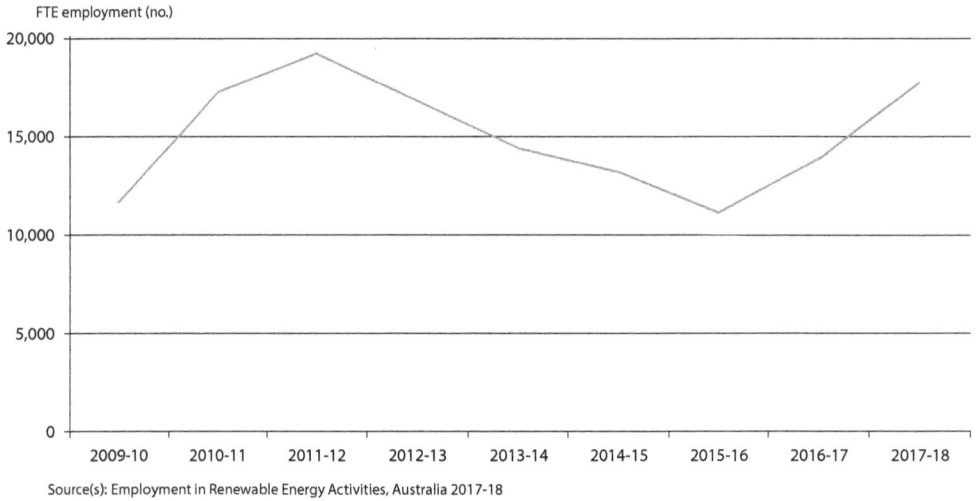

Source(s): Employment in Renewable Energy Activities, Australia 2017-18

Fig. 13.2. Annual direct full-time equivalent employment in renewable energy activities in Australia, 2009–10 to 2017–18. Source: ABS (2018).

influence. The release of a built-up pipeline early in the period is another, leading to an initial boost followed by a decline then a more gradual increase. Then in 2015–16, the construction of solar farms kicks in.

Over time, the contribution and sheer number of large-scale solar PVs have continued to grow, and that growth is projected to continue. The proportion of wind projects supporting employment has stayed relatively stable, although wind and solar projects have been found to be volatile as they are influenced by fluctuations in infrastructure capital formation. All states saw increases in renewable energy employment from 2012–13 to 2017–18. The patterns are somewhat irregular as a result of changes in feed-in tariffs and building of solar and wind farms.

FTE employment (no.)

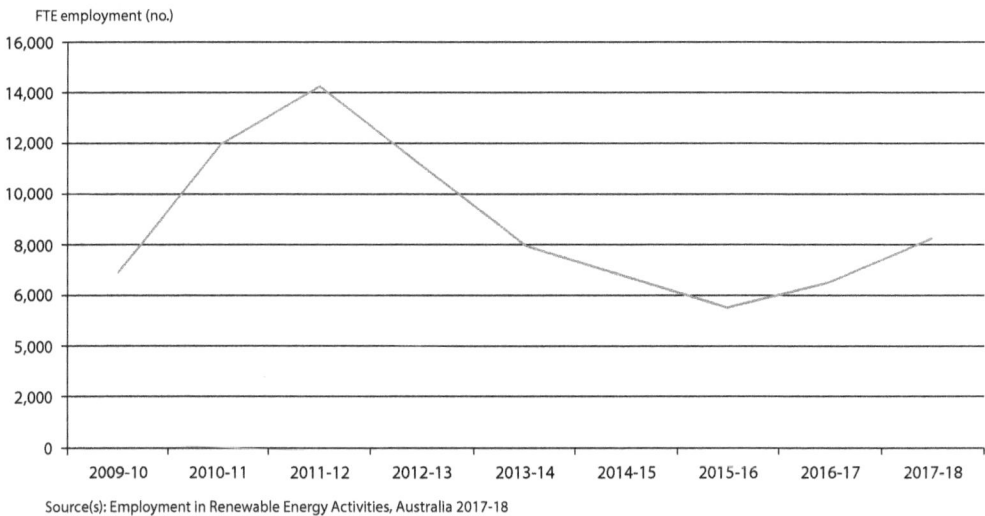

Source(s): Employment in Renewable Energy Activities, Australia 2017-18

Fig. 13.3. Annual direct full-time equivalent employment in rooftop solar activities in Australia, 2009–10 to 2017–18. Includes solar hot water systems. Source: ABS (2018).

The basic next steps in transition forecasts

One of the next steps is to forecast the installation of renewable energy sources over the coming decades, taking into account each type of energy generation and where in the nation it will take place. Various rates of development could be used to illustrate differing scenarios of the employment to be generated.

Another step is to construct a profile of the workers in the nation's coal mines. The most essential data are the number of workers according to skills and qualifications, their average wages plus entitlements, and whether they are on-site or FIFO (fly-in–fly-out) workers. All these data are available from the coal mine employers and should require no more than a handful of computer clicks. If, as could be the case, additional data are required (employee's age, number of dependants, home loans, etc.), a questionnaire-based survey should be administered to the workers.

We do not expect an ideal match between coal jobs and renewable energy jobs. Retraining of existing workers will be necessary, and more school leavers should be enticed to study for a trade. This will require another level of planning, which needs to start immediately.

As the data show, the number of jobs in the renewable energy industry is moving closer to those in coal mining. At this early stage of the renewable energy revolution, it provides half to three-quarters of the jobs in coal mining. At the rate of progress in the roll-out of solar and wind farms, plus the demand for workers in the Snowy 2.0 pumped-storage hydro-electricity project and battery-storage development, the coal mining industry will be quickly overtaken. This does not account for the need for workers in the soon-to-be-needed upgrade of the major electricity grid networks. Australians in the future will look back at the next two decades as the most significant revolution in energy production since fossil fuels replaced horse and human power.

References

ABS (2016) *3218.0 Regional Population Growth, Australia, 2014–15.* Australian Bureau of Statistics, Canberra. abs.gov.au

ABS (2018) *Employment in Renewable Energy Activities, Australia 2017–18.* Australian Bureau of Statistics, Canberra. CC BY 4.0.

AILA (2016) *Liveable Cities: Green Walls and Roofs.* Australian Institute of Landscape Architects, Canberra. aila.org.au

Allen GR, Midgley SH, Allen M (2002) *Field Guide to the Freshwater Fishes of Australia.* Western Australian Museum, Perth.

Allred SB, Ross-Davis A (2010) The drop-off and pick-up method: an approach to reduce nonresponse bias in natural resource surveys. *Small-scale Forestry* **9**, 305–318. doi:10.1007/s11842-010-9150-y

Amar JHN (2017) Conservation of Cultural Built Heritage: An Investigation of Stakeholder Perceptions in Australia and Tanzania. PhD thesis, Bond University, Robina.

AMBS (1999) *Water Reclamation and Management Scheme and Frog Habitat Works: Species Impact Statement.* Prepared for the Olympic Co-ordination Authority. Australian Museum Business Services, Sydney.

Angel O (2014) Green grow the cities. *Australian Geographic*, May–June, pp. 62–73. australiangeographic.com.au.

Anon (undated a) *Collins Dictionary.* HarperCollins, Sydney.

Anon (undated b) *A Call for a National Green Infrastructure Strategy.* Australian Institute of Landscape Architects, Canberra. aila.org.au

Archer D, Wearing S (2002) Interpretation and marketing as management tools in national parks: insights from Australia. *Journal of Retail & Leisure Property* **2**, 29–39. doi:10.1057/palgrave.rlp.5090137

Armstrong A (1986) Colonial and neo-colonial urban planning: three generations of master plans for Dar es Salaam, Tanzania. *Utafiti: Journal of the Faculty of Arts and Social Science* **8**, 43–66.

Ashton N (2016) *Learning to be Green: A Study of the Olympic Games.* MA, University of Ottawa, Ottawa.

Atfield C (2013) UN plans to list reef as endangered. *Sydney Morning Herald*, 4 May. smh.com.au/environment/un-plans-to-list-reef-as-endangered-20130504-2izkq.html

Balmford A, Beresford J, Green J, Naidoo R, Walpole M, Manica A (2009) A global perspective on trends in nature-based tourism. *PLoS Biology* **7**, e1000144. doi:10.1371/journal.pbio.1000144

Beder S (1993) Sydney's toxic green Olympics. *Current Affairs Bulletin* **70**, 12–18.

Beder S (1996) Sydney's Olympic landscape: a toxic coverup. *Search* **27**, 209–211.

Beeby R (2012) Replace Parliament's lawn with flowers, expert recommends. *Sydney Morning Herald*, 10 July. smh.com.au/environment/replace-parliaments-lawn-with-flowers-expert-recommends-20120710-21svb.html

Bishop K, Dudley N, Phillips A, Stolton S (2004) *Speaking a Common Language: The Uses and Performance of the IUCN System of Management Categories for Protected Lands.* World Conservation Union and Word Conservation Monitoring Centre, Paris.

Brennan J, Burton A, Lawi Y (2007) *The Emerging Metropolis: A Short History of Dar es Salaam, circa 1862–2005.* Mkuki Na Nyota/British Institute in Eastern Africa, Dar es Salaam.

Brockman CF (1962) The need for definitions. In *Proceedings of the 1st World Conference on National Parks.* (Ed. AB Adams) pp. 366–367. National Parks Service, Washington, DC.

Brooks TM, Mittermeier RA, da Fonseca GAB, Gerlach J, Hoffmann M, Lamoreaux JF, Mittermeier CG, Pilgrim JD, Rodrigues ASL (2006) Global biodiversity consideration priorities. *Science* **313**, 58–61. doi:10.1126/science.1127609

Buckley R (2009) Parks and tourism. *PLoS Biology* **7**, e1000143. doi:10.1371/journal.pbio.1000143

Bull JW, Suttle KB, Gordon A, Singh NJ (2013) Biodiversity offset in theory and practice. *Oryx* **47**, 369–380. doi:10.1017/S003060531200172X

Burgin S (2008) BioBanking: an environmental scientist's view of the role of biodiversity banking offsets in conservation. *Biodiversity and Conservation* **17**, 807–816. doi:10.1007/s10531-008-9319-2

Burgin S (2018a) 'Back to the future': transforming the urban landscape to support greater food self-sufficiency. *Land Use Policy* **78**, 29–35. doi:10.1016/j.landusepol.2018.06.012

Burgin S (2018b) Sustainability as a motive for leisure-time gardening: a view from the 'biggie patch'. *International Journal of Environmental Studies* **75**, 1000–1010. doi:10.1080/00207233.2

Burgin S, Hardiman D (2014) Maintaining competitive tourism advantage with reference to the Greater Blue Mountains World Heritage Area. In *Proceedings of the 7th International Conference on Monitoring and Management of Visitors in Recreational and Protected Areas*, 20–23 August 2014, Tallinn. (Eds M Reimann, K Sepp, E Parna, R Tuula), pp. 40–41.

Burgin S, Zama E (2014) Community-based tourism: option for forest-dependent communities in 1A IUCN protected areas? Cameroon case study. *SHS Web of Conferences* **12**, 01067.

Burgin S, Schell CB, Briggs C (2005) Is *Batrachochytrium dendrobatidis* really the proximate cause of frog decline? *Acta Zoologica Sinica* **5**, 344–348.

Burgin S, Mattila M, McPhee D, Hundloe T (2015) Feral deer in the suburbs: an emerging issue for Australia? *Human Dimensions of Wildlife* **20**, 65–80. doi:10.1080/10871209.2015.953274

Burgin S, Franklin M, Hull L (2016) Wetland loss in the transition to urbanisation: a case study from western Sydney. *Wetlands* **36**, 985–994. doi:10.1007/s13157-016-0813-0

Burns EL, Eldridge MDB, Houlden BA (2004) Microsatellite variation and population structure in a declining Australian hylid *Litoria aurea*. *Molecular Ecology* **13**, 1745–1757. doi:10.1111/j.1365-294X.2004.02190.x

Campbell N (2001) *Integrating ESD: The Environmental Legacy of Sydney Olympic Park*. Presentation at Environmental Symposium, 14 September 2001, Old Parliament House, Canberra. sopa.nsw.gov.au

Carey C, Dudley N, Stolton S (2000) *Threats to Protected Areas. Squandering Paradise? The Importance and Vulnerability of the World's Protected Areas*. World Wide Fund for Nature International, Gland.

Cashman R, Hughes A (Eds) (1998) *The Green Games: A Golden Opportunity*. Centre for Olympic Studies, University of New South Wales, Sydney.

Christy MT (2003) Making headlines or saving a species? *Australian Zoologist* **32**, 324–328. doi:10.7882/AZ.2003.016

Clark GF, Johnston EL (2016) Coasts: coastal governance and management. In *Australian State of the Environment 2016*. Department of the Environment and Energy, Canberra. doi:10.4226/94/58b659bdc758b

Clawson M, Knetsch J (1966) *Economics of Outdoor Recreation*. Johns Hopkins Press, Baltimore.

Cleary P (2011) *Too Much Luck: The Mining Boom and Australia's Future*. Black Inc, Melbourne.

Cogger HG (2014) *Reptiles and Amphibians of Australia*. 7th edn. CSIRO Publishing, Melbourne.

Colgan D (1996) Electrophoretic variation in the green and golden bell frog *Litoria aurea*. *Australian Zoologist* **30**, 170–176. doi:10.7882/AZ.1996.009

Commonwealth of Australia (2012) *Environment Protection and Biodiversity Conservation Act 1999: Environmental Offset Policy*. Department of Sustainability, Environment, Water, Population and Communities, Canberra.

Commonwealth of Australia (2017) *Policy Statement: Advanced Environmental Offsets Under the Environmental Protection and Biodiversity Conservation Act 1999*. Department of the Environment and Energy, Canberra. environment.gov.au

Conover DO, Munch SB (2002) Sustaining fisheries yields over evolutionary time scales. *Science* **297**, 94–96. doi:10.1126/science.1074085

Cunningham Bissell CW (2007) Casting a long shadow: colonial categories, cultural identities, and cosmopolitan spaces in globalizing Africa. *African Identities* **5**, 181–197. doi:10.1080/14725840701403416

Daly G, Johnson P, Malolakis G, Hyatt A, Pietsch R (2008) Reintroduction of the green and golden bell frog *Litoria aurea* to Pambula on the south coast of New South Wales. *Australian Zoologist* **34**, 261–270. doi:10.7882/AZ.2008.003

Darcovich K, O'Meara J (2008) An Olympic legacy: green and golden bell frog conservation at Sydney Olympic Park 1993–2006. *Australian Zoologist* **34**, 236–248. doi:10.7882/AZ.2008.001

Dasmann RF (1974) Development of a classification system for protected natural and cultural areas. In *Proceedings of the 2nd World Conference on National Parks*. Occasional Paper No. 7. (Ed. HB Elliott) IUCN, Morges.

DAWE (undated) *Protected Area Locations*. Department of Agriculture, Water and the Environment, Canberra. environment.gov.au

DECC (2005) *Green and Golden Bell Frog Litoria aurea (Lesson 1829) Recovery Plan*. NSW Department of Environment and Conservation – Recovery Plan Program. environment.nsw.gov.au

DECC (2008) *Management Plan: The Green and Golden Bell Frog Parramatta Key Population*. Department of Environment and Climate Change, Sydney.

DEHP (2016) *Annual Report 2015–2016: Department of Environment and Heritage Protection*. Qld Department of Environment and Heritage, Brisbane.

DEWHA (2008–09) *Annual Report*. Department of the Environment, Water, Heritage and the Arts, Canberra. environment.gov.au

Dillon M (2008) Sky high but down to earth. *Ecolibrium* **November**, 34–36.

DIRD (2013) *State of Australian Cities 2013*. Department of Infrastructure and Regional Development, Canberra.

DNPRSR (2014) *Annual Report 2013–2014*. Qld Department of National Parks, Recreation, Sport and Racing, Brisbane.

DNPSR (2017) *Vehicle Access Permits*. Qld Department of National Parks, Sport and Racing, Brisbane. npsr.qld.gov.au

DOE (2015) *Carmichael Coal Project – Frequently Asked Questions*. Department of the Environment, Canberra. environment.gov.au

Dowle M (2016) *Green and Golden Bell Frog Plan of Management – Arncliffe*. Eco Logical Australia, Sydney.

Dowling CE, Hall KC, Broadhurst MK (2010) Immediate fate of angled-and-released Australian bass *Macquaria novemaculeata*. *Hydrobiologia* **641**, 145–157. doi:10.1007/s10750-009-0073-6

DPI (2017) *Record Number of Australian Bass Stocked in NSW*. NSW Department of Primary Industries, Sydney. dpi.nsw.gov.au

DPI (undated) *Dollar for Dollar*. NSW Department of Primary Industries, Sydney. dpi.nsw.gov.au

Dudley N (2008) *Guidelines for Applying Protected Area Management Categories*. IUCN, Gland.

Dudley N, Higgins-Zogib L, Mansourian S (2005) *Beyond Belief: Linking Faiths and Protected Areas to Support Biodiversity Conservation*. World Wildlife Fund, Gland.

Dudley N, Parrish J (2006) *Closing the Gap: Creating Ecologically Representative Protected Area Systems*. Technical Services Issue 24. Secretariat of the Convention on Biological Diversity, Montreal.

Eagles PF, McCool SF (2002) *Tourism in National Parks and Protected Areas: Planning and Management*. CABI Publications, Wallingford.

Ebua VB, Agwafo TE, Fonkwo SN (2011) Attitudes and perceptions as threats to wildlife conservation in the Bakossi area, South West Cameroon. *International Journal of Biodiversity and Conservation* **3**, 631–636. doi:10.5897/IJBC

Elefante C (2007) The greenest building is … one that is already built. *Forum Journal* **21**, 26–38.

Elliott HB (Ed.) (1974) *Proceedings of the 2nd World Conference on National Parks*. IUCN, Morges.

EPA (undated) *Heat Island Effect*. US Environmental Protection Agency, Washington, DC. epa.gov

Erbes R (1973) *International Tourism and the Economy of Developing Countries*. OECD, Paris.

Fernando G, Metuge VE (2016) Impacts of Sustainable Tourism within South West Region of Cameroon. Bachelor thesis, Laurea University of Applied Sciences, Kerava.

Fletcher JJ (1889) Observations of the oviposition and habits of certain Australian batrachians. *Proceedings of the Linnean Society of New South Wales* **4**, 357–387. doi:10.5962/bhl.part.15049

Fonchingong TN (2010) Anglophone Cameroon in search of indent: obstacles and prospects. *Africa Insight* **40**, 19–38.

Freese B (2003) *Coal: A Human History*. Basic Books, New York.

GBRMPA (2014) *Great Barrier Reef Outlook Report 2014*. Great Barrier Reef Marine Park Authority, Townsville. http://hdl.handle.net/11017/2855

Gibbons P, Lindenmayer DB (2007) Offsets for land clearing: no net loss or the tail wagging the dog. *Ecological Management & Restoration* **8**, 26–31. doi:10.1111/j.1442-8903.2007.00328.x

Gibbons P, Evans MC, Maron M, Gordon A, Le Roux D, et al. (2016) A loss–gain calculator for biodiversity offsets and the circumstances in which no net loss is feasible. *Conservation Letters* **9**, 252–259. doi:10.1111/conl.12206

Gill SE, Handley JF, Ennos R, Pauleit S (2007) Adapting cities for climate change: the role of the green infrastructure. *Built Environment* **33**, 115–133. doi:10.2148/benv.33.1.115

Gillespie GR (1996) Distribution, habitat and conservation status of the green and golden bell frog *Litoria aurea* (Lesson 1829) (Anura: Hylidae) in Victoria. *Australian Zoologist* **30**, 199–207. doi:10.7882/AZ.1996.012

Gledhill H (2011) *Victorian Desalination Project Coastal Park*. depi.vic.gov.au/__data/assets/pdf_file/0005/190481/ASPECT-Studios-presentation-May-2011.pdf

Goldingay RL (1996) The green and golden bell frog (*Litoria aurea*): from riches to ruins – conservation of a formerly common species. *Australian Zoologist* **30**, 248–256. doi:10.7882/AZ.1996.019

Goldingay R, Osborne W (Eds) (2008) *Ecology and Conservation of Australian Bell Frogs. Australian Zoologist* **34**(3).

Goldman HV (2014) Tourism and social change in post-socialist Zanzibar: struggles for identity, movement, and civilization. *Journal of Modern African Studies* **52**, 505–507. doi:10.1017/S0022278X14000330

Gray CA, McDonall VC, Reid DD (1990) Bycatch from prawn trawling in the Hawkesbury River, New South Wales: species composition, distribution and abundance. *Australian Journal of Marine and Freshwater Research* **41**, 13–26. doi:10.1071/MF9900013

Gray CA, Kennelly SJ, Hodgson KE (2003) Low levels of bycatch from estuarine prawn seining in New South Wales, Australia. *Fisheries Research* **64**, 37–54. doi:10.1016/S0165-7836(03)00185-1

Greer AE (1994) *Faunal Impact Statement for the Proposed Development Works at the Homebush Bay Brickpit*. Prepared for the Property Services Group, Sydney.

Greer AE (1996) Why green and golden bell frog *Litoria aurea* should not be translocated: a personal opinion. *Australian Zoologist* **30**, 257–258. doi:10.7882/AZ.1996.020

Growns I, James M (2005) Relationships between river flows and recreational catches of Australian bass. *Journal of Fish Biology* **66**, 404–416. doi:10.1111/j.0022-1112.2005.00605.x

Haji HA, Azzan RM, Ufuzo SS (2006) Evolution of spatial planning in Zanzibar and its influence. Paper presented at the *Shaping the Change XXIII FIG Congress*, 8–13 October, Munich.

Hamer A, Lane S, Mahony M (2002) The role of introduced mosquitofish (*Gambusia holbrooki*) in excluding the native green and golden bell frog (*Litoria aurea*) from original habitats in south-eastern Australia. *Oecologia* **132**, 445–452. doi:10.1007/s00442-002-0968-7

Hansen AJ, DeFries R (2007) Ecological mechanisms linking protected areas to surrounding lands. *Ecological Applications* **17**, 974–988. doi:10.1890/05-1098

Hardiman N, Burgin S (2013) World Heritage Area listing on the Greater Blue Mountains: did it make a difference to visitation? *Tourism Management Perspectives* **6**, 63–64. doi:10.1016/j.tmp.2012.12.002

Harris JH (1983) The Australian Bass, *Macquaria novemaculeata*. PhD thesis, University of New South Wales, Sydney.

Harris JH (1985) Age of the Australian bass, *Macquaria novemaculeata* (Perciformes: Percichthyidae), in the Sydney Basin. *Australian Journal of Marine and Freshwater Research* **36**, 235–246. doi:10.1071/MF9850235

Harris JH (1986) Reproduction of the Australian bass, *Macquaria novemaculeata* (Perciformes: Percichthyidae), in the Sydney Basin. *Australian Journal of Marine and Freshwater Research* **37**, 209–235. doi:10.1071/MF9860209

Harris JH (1988) Demography of Australian bass, *Macquaria novemaculeata* (Perciformes: Percichthyidae), in the Sydney Basin. *Australian Journal of Marine and Freshwater Research* **39**, 355–369. doi:10.1071/MF9880355

Hatfield D, VanderKooy J, Bleeker M (2002) *A Sociolinguistic Survey among the Bakossi*. Summer Institute of Linguistics, Dallas.

Helder M, Chen WS, van der Harst EJM, Strik DPBTB, Hamelers HVM, Buisman CJN, Potting J (2013) Electricity production with living plants on a green roof: environmental performance of the plant-microbial fuel cell. *Biofuels, Bioproducts & Biorefining* **7**, 52–64. doi:10.1002/bbb.1373

Henry GW, Lyle JM (2003) *The National Recreational and Indigenous Fishing Survey*. Department of Agriculture, Fisheries and Forestry, Canberra.

Hill R, Miller C, Newell B, Dunlop M, Gordon IJ (2015) Why biodiversity declines as protected areas increase: the effect of the power of governance regimes on sustainable landscapes. *Natural Hazards* **10**, 357–369. doi:10.1007/s11652-05-0288-6

Hitchcock M (2002) Zanzibar Stone Town joins the imagined community of World Heritage sites. *International Journal of Heritage Studies* **8**, 153–166. doi:10.1080/13527250220143931

Holdgate M (1999) *The Green Web: A Union for World Conservation*. Earthscan Publications, London.

Hoyle B (2002) Port-city renewal in developing countries: the waterfront at Dar es Salaam, Tanzania.

Erdkunde **56**, 114–129. doi:10.3112/erdkunde.2002.02.01

Hundloe T (2018) *Adani Versus the Black-Throated Finch*. Australian Scholarly Publishing, Melbourne.

Hussein J, Armitage L (2014) Traditional heritage management: the case of Australia and Tanzania. In *Proceedings of the Asian Real Estate Society 19th Annual Conference*, 14–16 July, Gold Coast, Australia.

Hutton D, Conners L (1999) *A History of the Australian Environment Movement*. Cambridge University Press, Cambridge.

Ichumbaki E (2012) The State, Cultural Significance and Management of Built Heritage Assets of Lindi and Mtwara Regions, Tanzania. MA dissertation, University of Dar es Salaam, Dar es Salaam.

ICMM (2005) *Biodiversity Offsets: A Briefing Paper for the Mining Industry*. International Council on Mining and Metals, London.

IUCN (1994) *Guidelines for Protected Area Management Categories*. International Union for the Conservation of Nature, Gland.

IUCN (2012) *IUCN Red List Categories and Criteria: Version 3.1*. 2nd edn. International Union for the Conservation of Nature, Gland.

IUCN (2016) *The IUCN Red List of Threatened Species: Summary Statistics*. International Union for the Conservation of Nature, Gland. iucnredlist.org

IUCN (2017) *Protected Areas*. International Union for the Conservation of Nature, Gland. iucn.org

Jacobs J (2001) Æsop Fables (retold). Vol. 17, The Harvard Classics (1909–1914). PF Collier and Son, New York. bartleby.com/17/1/

Jenkins CN, Joppa L (2009) Expansion of the global terrestrial protected area system. *Biological Conservation* **142**, 2166–2174. doi:10.1016/j.biocon.2009.04.016

Kasala ES (2015) A return to master planning: a misconception of the theory of paradigm shift? *Global Journal of Human-Social Science* **15**, 1–7.

Khalfan KA (2014) *Waqf* as a model for production and conservation of architectural heritage. *Global Journal of Human-Social Science* **14**, 19–26.

Khalfan KA, Ogura N (2010) Influence of outsourced finance on conservation of built heritage in a developing country: the case of Stone Town of Zanzibar, Tanzania. *Journal of Architecture and Planning* **75**, 2507–2515. doi:10.3130/aija.75.2507

Khalfan KA, Ogura N (2012) Sustainable architectural conservation according to traditions of Islamic waqf: the World Heritage-listed Stone Town of Zanzibar. *International Journal of Heritage Studies* **18**, 588–604. doi:10.1080/13527258.2011.607175

Komu F (2011) Housing Decay and Maintenance: The Case of Public Housing in Tanzania. PhD thesis, KTH Royal Institute of Technology, Stockholm.

Kongela SM (2013) Framework and Value Drivers for Real Estate Development in Sub-Saharan Africa: Assessment of the Tanzanian Real Estate Sector in the Context of the Competitiveness Model. PhD thesis, University of Regensburg, Regensburg.

Lall SV, Freire M, Yuen B, Rajack R, Helluin JJ (2009) *Urban Land Markets: Improving Land Management for Successful Urbanization.* Springer Science, London.

Lambi CM, Ndenecho EN (2009) *Ecology and Natural Resource Development in the Western Highlands of Cameroon: Issues in Natural Resource Management.* African Books Collective, Oxford.

Land Court of Queensland (2015) *Adani Mining Pty Ltd v Land Services of Coast and Country Inc & Ors* [2015] QLC 48.

Lemckert F (2010) The rich early history of frog research in Sydney. In *The Natural History of Sydney.* (Eds D Lunney, P Hutchings, D Hochuli) pp. 102–106. Royal Zoological Society of New South Wales, Sydney.

Lenskyj HJ (1998) Sport and corporate environmentalism: the case of the Sydney 2000 Olympics. *International Review for the Sociology of Sport* 33, 341–354. doi:10.1177/101269098033004002

Lenskyj HJ (2000) *Inside the Olympic Industry: Power, Politics, and Activism.* State University of New York Press, Albany.

Leopold A (1990) *Sand County Almanac: And Sketches Here and There.* Oxford University Press, New York.

Loh S (2009) Green roofs: understanding their benefits for Australia. *Environment Design Guide* 27, 1–13.

Lunney D, Ayers D (1993) The official status of frogs and reptiles in New South Wales. In *Herpetology in Australia: A Diverse Discipline.* (Eds D Lunney, D Ayers) pp. 404–408. Transactions of the Royal Zoological Society of New South Wales, Surrey Beatty and Sons, Sydney.

Lunney D, Curtin A, Ayers D, Cogger HG, Dickman CR (1995) An ecological approach to identifying the endangered fauna of New South Wales. *Pacific Conservation Biology* 2, 212–231. doi:10.1071/PC960212

Lwoga NB (2013) Tourism development in Tanzania before and after independence: sustainability perspectives. *Eastern African Journal of Hospitality, Leisure and Tourism* 1, 1–26.

Mahony M (1996) The decline of the green and golden bell frog *Litoria aurea* viewed in the context of declines and disappearances of other Australian frogs. *Australian Zoology* 30, 237–247. doi:10.7882/AZ.1996.018

Mahony MJ, Hamer AJ, Pickett EJ, McKenzie DJ, Stockwell MP, Garnham JI, Keely CC, Deboo ML, O'Meara J, Pollard CJ, Clulow S, Lemckert FL, Bower DS, Clulow J (2013) Identifying conservation and research priorities in the face of uncertainty: a review of the threatened bell frog complex in eastern Australia. *Herpetological Conservation and Biology* 8, 519–538.

Margules CR, Pressey RL (2000) Systematic conservation planning. *Nature* 405, 243–253. doi:10.1038/35012251

MRC (undated) *Australia's Marine Protected Area Network.* Marine Reserves Coalition, UK. marinereservescoalition.org

Marks R (1996) Conservation and community: the contradictions and ambiguities of tourism in the Stone Town of Zanzibar. *Habitat International* 20, 265–278. doi:10.1016/0197-3975(95)00062-3

Maron M, Gordon A, Mackey BG, Possingham HP, Watson JEM (2015) Stop misuse of biodiversity offsets. *Nature* 523, 401–403. doi:10.1038/523401a

Maron M, Gordon A, Mackey BG (2016) Interactions between biodiversity offsets and protected area commitments: avoiding perverse outcomes. *Conservation Letters* 9, 384–389. doi:10.1111/conl.12222

Masele F (2012) Private business investments in heritage sites in Tanzania: recent developments and challenges for heritage management. *African Archaeological Review* 29, 51–65. doi:10.1007/s10437-012-9105-0

McAlpine C, Seabrook L (undated) *The Brigalow.* In *Queensland Historical Atlas: Histories, Cultures and Landscapes.* qhatlas.com.au/content/brigalow

McFadden M, Duffy S, Harlow P, Hobcroft D, Webb C, Ward-Fear G (2008) A review of the green and golden bell frog *Litoria aurea* breeding program at Taronga Zoo. *Australian Zoologist* 34, 291–296. doi:10.7882/AZ.2008.006

McGeoch MA, Dopolo M, Novellie P, Hendriks H, Freitag S, Ferreira S, Grant R, Kruger J, Bezuidenhout H, Randall RM, Vermeulen W, Kraaij T, Russell IA, Knight MH, Holness S, Oosthuizen A (2011) A strategic framework for biodiversity monitoring in South African national parks. *Koedoe* 53, 43–51. doi:10.4102/koedoe.v53i2.991

Menzies NK (1992) Strategic space: exclusion and inclusion in wildland policies in late Imperial China. *Modern Asian Studies* 26, 719–733. doi:10.1017/S0026749X00010040

Midgley A (2015) Ecology of the Australian Bass in Tributaries of the Hawkesbury-Nepean River. PhD thesis, Western Sydney University, Penrith.

Miehs A, Pyke GH (2001) Observations of the foraging behaviour of adult green and golden bell frogs (*Litoria aurea*). *Herpetofauna* 31, 94–96.

MNRT (2014) *An Overview of the Antiquities Sub Sector: Achievements, Challenges and Priorities for Financial Year 2014/15.* A paper presented by Mr Donatius Kamamba, Director for Antiquities at the 2014 Natural Resources Sector Review Meeting on 16 October 2014 at the National College of Tourism, Bustani Compus-Dar es Salaam. Ministry of Natural Resources and Tourism, Dar es Salaam, Tanzania.

Mokany A, Shine R (2003) Competition between tadpoles and mosquito larvae. *Oecologia* 135, 615–620. doi:10.1007/s00442-003-1215-6

Monty F, Murti R, Miththapala S, Buyk C (2017) *Ecosystems Protecting Infrastructure and Communities:*

Lessons Learned and Guidelines for Implementation. IUCN, Gland.

Naughton-Treves L, Holland MB, Brandon K (2005) The role of protected areas in conserving biodiversity and sustaining local livelihoods. *Annual Review of Environment and Resources* **30**, 219–252. doi:10.1146/annurev.energy.30.050504.164507

NBS (2011) *Measuring and Valuing Environmental Impacts: An Introductory Guide.* Network for Business Sustainability, Ontario.

Ngea P (2011) *Mount Kupe and Muanenguba: Custodian of Tradition and Biodiversity.* World Wide Fund for Nature, Washington, DC.

Nnkya TJ (2007) *Why Planning does not Work: Land Use Planning and Residents' Rights in Tanzania.* Mkuki Na Nyota Publishers, Dar es Salaam.

Norton B, Bosomworth K, Coutts A, Williams N, Livesley S, Trundle A, Harris R, McEvoy D (2013) *Planning for a Cooler Future: Green Infrastructure to Reduce Urban Heat.* Final Report. Victorian Centre for Climate Change Adaptation Research, Melbourne.

Norton BA, Coutts AM, Livesley SJ, Harris RJ, Hunter AM, Williams NSG (2015) Planning for cooler cities: a framework to prioritise green infrastructure to mitigate high temperatures in urban landscapes. *Landscape and Urban Planning* **134**, 127–138. doi:10.1016/j.landurbplan.2014.10.018

NSC (2020) *Coal Loader Platform Green Roof*: Coal Loader Centre for Sustainability, North Sydney Council, Sydney. northsydney.nsw.gov.au

NSW Govt (2017) *Annual Entry Fees.* NSW National Parks and Wildlife Service, Sydney. nationalparks. nsw.gov.au

O'Meara J, Darcovich K (2015) Twelve years on: ecological restoration and rehabilitation at Sydney Olympic Park. *Ecological Management & Restoration* **16**, 14–28. doi:10.1111/emr.12150

Oberauer N (2008) 'Fantastic charities': the transformation of *waqf* practice in colonial Zanzibar. *Islamic Law and Society* **15**, 315–370. doi:10.1163/156851908X366156

Oberndorfer E, Lundholm J, Bass B (2007) Green roofs as urban ecosystems: ecological structures, functions, and services. *Bioscience* **57**, 823–833. doi:10.1641/B571005

OEH (2015) *Threatened Species: Green and Golden Bell Frog: Profile.* Office of Environment and Heritage, Sydney. nsw.gov.au.

Olsen S, Amundsen DA, Anderson D, Guy S (1998) Community interest survey to plan Utah Botanical Center. *Journal of Extension* **36**, 6TOT2.

Osborne WS, Littlejohn MJ, Thomson SA (1996) Former distribution and apparent disappearance of the *Litoria aurea* complex from the Southern Tablelands of New South Wales and the Australian Capital Territory. *Australian Zoologist* **30**, 190–198. doi:10.7882/AZ.1996.011

Owens K (2012) Enabling sustainable markets? The redevelopment of Dar es Salaam. Paper presented at *the World Bank's 6th Urban Research and Knowledge Symposium,* 8–10 October, Barcelona. worldbank.org

Parks Australia (2020) *Australian Marine Parks.* Parks Australia, Canberra. parksaustralia.gov

Penman TD, Muir GW, Magarey ER, Burns EL (2008) Impact of a chytrid-related mortality event on a population of the green and golden bell frog *Litoria aurea. Australian Zoologist* **34**, 314–318. doi:10.7882/AZ.2008.009

Phillips A (2004) Protected area categories. *Parks* **14**, 4–13.

Pickett EJ, Stockwell MP, Bower DS, Garnham JI, Pollard CJ, Clulow J, Mahony MJ (2013) Achieving no net loss in habitat offset of a threatened frog required high offset ratio and intensive monitoring. *Biological Conservation* **157**, 156–162. doi:10.1016/j.biocon.2012.09.014

Pickett EJ, Stockwell MP, Bower DS, Garnham JI, Pollard CJ, Garnham JI, Clulow J, Mahony MJ (2014) Six-year demographic study reveals threat of stochastic extinction of green and golden bell frog tadpoles. *Herpetological Journal* **26**, 157–164. doi:10.1111/aec.12080

Planning Commission (1999) *The Tanzania Development Vision 2025.* President's Office, Dar es Salaam.

Pollard DA, Growns IO (1993) *The Fish and Fisheries of the Hawkesbury-Nepean River Sydney, with Particular Reference to the Environmental Effect of the Water Board's Activities on this System.* NSW Fisheries Institute, Sydney.

Pounds JA, Bustamante MR, Coloma LA, Consuegra JA, Fogden MPF, Foster PN, La Marca E, Masters KL, Merino-Viteri A, Puschendorf R, Ron SR, Sánchez-Azofeifa GA, Still CJ, Young BE (2006) Widespread amphibian extinctions from epidemic disease driven by global warming. *Nature* **439**, 161–167. doi:10.1038/nature04246

Pyke G, Miehs A (2001) Predation by water skinks (*Eulamprus quoyii*) on tadpoles and metamorphs of the green and golden bell frog (*Litoria aurea*). *Herpetofauna* **31**, 99–101.

Pyke GH, Osborne WS (Eds) (1996) *The Green and Golden Bell Frog Litoria aurea: Biology and Conservation. Australian Zoologist* **30**(2).

Pyke G, White A (1996) Habitat requirements for the green and golden bell frog *Litoria aurea* (Anura: Hylidae). *Australian Zoologist* **30**, 224–232.

Pyke G, White A (2000) Factors influencing predation on eggs and tadpoles of the green and golden bell frog *Litoria aurea* by the introduced plague minnow *Gambusia holbrooki. Australian Zoologist* **31**, 496–505. doi:10.7882/AZ.2000.011

Pyke G, White A (2001) A review of the biology of the green and golden bell frog *Litoria aurea. Australian Zoologist* **31**, 563–598. doi:10.7882/AZ.2001.003

Pyke G, White A, Bishop PJ, Waldman B (2002) Habitat-use by the green and golden bell frog *Litoria aurea* in Australia and New Zealand. *Australian Zoologist* **32**, 12–31. doi:10.7882/AZ.2002.002

Pyke GH, Rowley J, Shoulder J, White AW (2008) Attempted introduction of the endangered green and golden bell frog to Long Reef Golf Course: a step towards recovery? *Australian Zoologist* **34**, 361–372. doi:10.7882/AZ.2008.013

Pyke GH, Ahyong ST, Fuessel A, Callaghan S (2013) Marine crabs eating freshwater frogs: why are such observations so rare? *Herpetology Notes* **6**, 195–199.

Razzaghmanesh M, Beecham S, Brien CJ (2014) Developing resilient green roofs in a dry climate. *Science of the Total Environment* **490**, 579–589. doi:10.1016/j.scitotenv.2014.05.040

Recfish West (2014) *Position Statement: Recognising the Values of Recreational Fishing.* WA Department of Primary Industries and Regional Development (Fisheries), Perth. recfishwest.org.au

Redford KH, Fearn E (2007) *Protected Areas and Human Livelihoods.* Working Paper No. 32. Wildlife Conservation Society, New York.

Rhodes DT (2014) *Building Colonialism: Archaeology and Urban Space in East Africa.* Bloomsbury Publishing, London.

Robertson G (1993) Foreword. In *Reintegrating Fragmented Landscapes: Towards Sustainable Production and Nature Conservation.* (Eds RJ Hobbs, DA Saunders) pp. vii–x. Springer-Verlag, New York.

Rosenzweig C, Solecki W, Slosberg R (2006) *Mitigating New York City's Heat Island with Urban Forestry, Living Roofs, and Light Surfaces.* giss.nasa.gov/research/news/20060130/103341.pdf

Seabrook L, McAlpine C, Maron M (2016) EcoCheck: can the Brigalow Belt bounce back? Environment + Energy. *The Conversation*, 6 May. theconversation.com

Sellars RW (1997) *Preserving Nature in the National Parks: A History.* Yale University Press, New Haven.

Shaddick K, Burridge CP, Jerry DR, Schwartz TS, Troung K, Gilligan DM, Beheregaray LB (2011) A hybrid zone and bidirectional introgression between two catadromous species: Australian bass *Macquaria novemaculeata* and estuary perch *Macquaria colonorum. Journal of Fish Biology* **79**, 1214–1235. doi:10.1111/j.1095-8649.2011.03105.x

Sheriff A (1995) *The History and Conservation of Zanzibar Stone Town.* Department of Archives, Museums and Antiquities Zanzibar in association with J Currey. Ohio University Press, Athens.

Sheriff A (2001) *Zanzibar Stone Town: An Architectural Exploration.* The Gallery Publications, Zanzibar.

Shuster WD, Morrison MA, Webb R (2008) Front-loading urban stormwater management for success: a perspective incorporating current studies on the implementation of retrofit low-impact develop-ment. *Cities and the Environment* **1**(2), 8. doi:10.15365/cate.1282008

Siravo F (1996) *Zanzibar: A Plan for the Historic Stone Town.* The Gallery Publications, Zanzibar.

Skerratt LF, Berger L, Speare R, Cashins S, McDonald K, Phillott R, Dawn A, Hines HB, Kenyon N (2007) Spread of chytridiomycosis has caused the rapid global decline and extinction of frogs. *EcoHealth* **4**, 125. doi:10.1007/s10393-007-0093-5

SOPA (2006) *Annual Report 2005–06.* Sydney Olympic Park Authority, Sydney.

SOPA (2007) *State of the Environment Report 2006–2007.* Sydney Olympic Park Authority, Sydney.

SOPA (2008) *Environmental Guidelines Sydney Olympic Park 2008.* Sydney Olympic Park Authority, Sydney.

SOPA (2009) *Annual Report 2008–09.* Sydney Olympic Park Authority, Sydney.

SOPA (2011) *State of the Environment Report 2010–11.* Sydney Olympic Park Authority, Sydney.

SOPA (2012) *Sydney Olympic Park Authority Annual Report 2011–2012.* Sydney Olympic Park Authority, Sydney.

SOPA (2014) *Conserving the Green and Golden Bell Frogs.* Sydney Olympic Park Authority, Sydney.

SOPA (2015) *Sydney Olympic Park Authority Annual Report: 2014–2015.* Sydney Olympic Park Authority, Sydney.

SOPA (2016) *Sydney Olympic Park Authority Annual Report: 2015–2016.* Sydney Olympic Park Authority, Sydney.

Spenceley A (2012) *Responsible Tourism: Critical Issues for Conservation and Development.* Routledge, London.

State of Queensland (2015) *Nature Conservation Act 1992.* legislation.qld.gov.au/LEGISLTN/CURRENT/N/NatureConA92.pdf

Steffen W (2015) *Galilee Basin: Unburnable Coal.* Climate Change Council of Australia, Sydney. climatecouncil.org.au

Stocks M, Blakers A (2018) New coal doesn't stack up: just look at Queensland's renewable energy numbers. *The Conversation*, 27 June. theconversation.com

Stockwell MP, Clulow S, Clulow J, Mahony M (2008) The impact of the amphibian chytrid fungus *Batrachochytrium dendrobatidis* on a green and golden bell frog *Litoria aurea* reintroduction program at the Hunter Wetlands Centre Australia in the Hunter Region of NSW. *Australian Zoologist* **34**, 379–386. doi:10.7882/AZ.2008.015

Stockwell MP, Clulow J, Mahony MJ (2010) Host species determines whether infection load increases beyond disease-causing thresholds following exposure to the amphibian chytrid fungus. *Animal Conservation* **13**, 62–71. doi:10.1111/j.1469-1795.2010.00407.x

Stockwell MP, Bower DS, Bainbridge L, Clulow J, Mahony MJ (2015) Island provides a pathogen refugee within climatically suitable area. *Biodiversity and Conservation* **24**, 2583–2592. doi:10.1007/s10531-015-0946-0

Stuart SN, Chanson JS, Cox NA, Young BE, Rodrigues ASL, Fischman DL, Waller RW (2004) Status and trends of amphibian declines and extinctions worldwide. *Science* **306**, 1783–1786. doi:10.1126/science.1103538

Threlfall CG, Jolley DF, Evershed N, Goldingay RL, Buttemer WA (2008) Do green and golden bell frogs *Litoria aurea* occupy habitats with fungicidal properties? *Australian Zoologist* **34**, 350–360. doi:10.7882/AZ.2008.012

Thurley T, Bell BD (1994) Habitat distribution and predation on a western population of terrestrial *Leiopelma* (Anura: Leiopelmatidae) in the northern King Country, New Zealand. *New Zealand Journal of Zoology* **21**, 431–436. doi:10.1080/03014223.1994.9518013

UNDP (2016) *Human Development Report 2016: Human Development for Everyone.* UN Development Program. hdr.undp.org

UNEP–WCMC, IUCN (2016) *Protected Planet Report 2016.* UN Environment World Conservation Monitoring Programme and International Union for Conservation of Nature, Cambridge and Gland.

UNEP–WCMC, IUCN (2018) *Protected Planet Report 2018.* UN Environment World Conservation Monitoring Programme and International Union for Conservation of Nature, Cambridge and Gland.

UN (2017) *World Population Prospects: The 2017 Revision.* UN Department of Economics and Social Affairs. un.org

Virtue R (2017) *NSW Scientists Breed Record 429,000 Australian Bass in Bumper Year for Fish Restocking Program. ABC News,* 19 October. abc.net.au

Watson JEM, Dudley N, Segan DB, Hockings M (2014) The performance and potential of protected areas. *Nature* **515**, 67–73. doi:10.1038/nature13947

Wetser K, Sudirjo E, Buisman CJN, Strik DPBTB (2015) Electricity generation by a plant microbial fuel cell with an integrated oxygen-reducing biocathode. *Applied Energy* **137**, 151–157. doi:10.1016/j.apenergy.2014.10.006

White AW (1998) *Compensatory Habitat, Monitoring and Long-term Maintenance of the Marsh Street Wetlands' Green and Golden Bell Frogs.* M5 East Motorway Project. Prepared for the NSW Roads and Traffic Authority, Sydney.

White AW, Burgin S (2004) Current status and future prospects of reptiles and frogs in Sydney's urban-impacted bushland reserves. In *Urban Wildlife: More than Meets the Eye.* (Eds D Lunney, S Burgin) pp. 109–123. Royal Zoological Society of New South Wales, Sydney.

White AW, Pyke GH (1996) Distribution and conservation of the green and golden bell frog *Litoria aurea* in New South Wales. *Australian Zoologist* **30**, 177–189. doi:10.7882/AZ.1996.010

White AW, Pyke GH (2008a) Frogs on the hop: translocations of green and golden bell frogs *Litoria aurea* in Greater Sydney. *Australian Zoologist* **34**, 249–260. doi:10.7882/AZ.2008.002

White AW, Pyke GH (2008b) Green and golden bell frogs in New South Wales: current status and future prospects. *Australian Zoologist* **34**, 319–333. doi:10.7882/AZ.2008.010

Wilde GR, Sawynok W (2005) Growth rate and mortality of Australian bass, *Macquaria novemaculeata,* in four freshwater impoundments in south-eastern Queensland, Australia. *Fisheries Management and Ecology* **12**, 1–7. doi:10.1111/j.1365-2400.2004.00412.x

Williams NSG, Rayner JP, Raynor KJ (2010) Green roofs for a wide brown land: opportunities and barriers for rooftop greening in Australia. *Urban Forestry and Urban Greening* **9**, 245–251. doi:10.1016/j.ufug.2010.01.005

Woods LJ, Fish L, Laughren J, Pauly D (2008) Assessing progress towards global marine protection targets: shortfalls in information and action. *Oryx* **42**, 340–351. doi:10.1017/S00306053080046X

World Bank (2015) *Tanzania's Tourism Futures Harnessing Natural Assets: Environment and Natural Resources Global Practice Policy Note.* World Bank Group, Washington, DC.

World Bank (2017) *Solar Powers India's Clean Energy Revolution,* 29 June 2017. World Bank Group, Washington, DC.

Wotherspoon D, Burgin S (2011) The impact on native herpetofauna due to traffic collision at the interface between a suburban area and the Greater Blue Mountains World Heritage Area: an ecological disaster? *Australian Zoologist* **35**, 1040–1046. doi:10.7882/AZ.2011.059

Yahya S (2013) *Reconciling Conservation Objectives with Development Imperatives: Lessons from the Zanzibar Stone Town Heritage Management Plan.* Paper presented at INTO Conference, 30 September–4 October, Entebbe. crossculturalfoundation.or.ug

Zahniser H (1992) *Where Wilderness Preservation Began: Adirondack Writings of Howard Zahniser.* North Country Books, Utica.

Index